PHYSICS AND COMPUTERS
Problems, Simulations, and Data Analysis

ROBERT EHRLICH
State University College
New Paltz, New York

HOUGHTON MIFFLIN COMPANY   BOSTON
Atlanta   Dallas   Geneva, Illinois   Hopewell, New Jersey   Palo Alto

To Elaine, David, and Gary

Printed in the United States of America

Library of Congress Catalog Card Number: 72-5640

ISBN: 0-395-18010-4

Preface

This text illustrates a number of computer applications in physics at the intermediate undergraduate level. Much of the material has been used in a three-credit, one-semester course entitled "Physics and Computers." The prerequisites for this course are one semester of calculus and one semester of general physics. This author feels that there is both a great need and a significant demand for new courses of this kind, and further, that such courses should be taught within the physics curriculum.

This book can also be used to supplement existing physics courses. The topics covered comprise a set of thirteen related, yet independent, problems selected from the areas of electricity, waves, and modern physics (see Table 1.1 on page 2). As a guide, Appendix I lists cross references between the thirteen topics listed in Table 1.1 and twenty-four commonly used introductory physics texts. The particular set of topics was chosen because they illustrate a wide variety of numerical methods and computer techniques: numerical integration, numerical solution of differential equations, computer graphics, random number generation, least-squares fitting. Further, many of these involve simulations of experiments that are often done in the physics laboratory: equipotential plotting, RC and RLC circuits, interference of waves from two sources, analysis of bubble chamber photographs. These computer methods enable the student to deal with problems that are too complex to treat using other methods. The simulated laboratory experiments are helpful in showing the student how the results of an experiment depend on a number of controllable parameters. Thus, the book can function as a partial alternative to the physics laboratory.

The first chapter consists of an introduction to computers, flow charts, and the FORTRAN IV language; it is not necessary (though it would be helpful) that the reader have previous experience with computers or the FORTRAN IV language. In the remaining four chapters each of the topics listed in Table 1.1 is discussed in the text and illustrated by completely documented computer programs. All the computer programs are written in FORTRAN IV and have been run on an 8000-word IBM 1130 computer. The ample documentation and flow charts should make the programs relatively language and computer independent. Those few computer-dependent features of the programs are discussed in Appendix II.

At the end of each chapter there is a substantial set of unworked problems, most of which are quite demanding. Some of the problems involve running the "canned" programs given in the chapter using various values for a number of input parameters; others involve some modification of the canned programs; still others involve writing completely new programs. In view of the relative independence of each topic, it is desirable that problems be worked after the completion of individual topics; thus the problems at the end of the chapters are subdivided according to topic within the chapter.

I want to express my gratitude to Dr. Joseph T. Ratau, Chairman of the Physics Department at New Paltz, and to the other members of the Physics Department, particularly Dr. Allan Harkavy, for their suggestions and for their approval and encouragement of the course "Physics and Computers." I would also like to thank those students who have taken this course for their comments and suggestions. In particular, I would like to gratefully acknowledge the many useful criticisms made by Mr. William F. Rall who painstakingly went through two versions of the manuscript. I am especially grateful to Dr. Ronald Blum who also read the entire manuscript and made numerous helpful comments and criticisms.

I am indebted to the Literary Executor of the late Sir Ronald A. Fisher, F.R.S., and to Oliver & Boyd, Edinburgh, for their permission to reprint Table III from their book *Statistical Methods for Research Workers*.

Robert Ehrlich

Table of Contents

1

Computers, Physics and FORTRAN

In recent years much progress in physics and technology has been made possible through the use of the high speed digital computer. The topics in this book illustrate some of the ways that the computer can prove useful in physics at the college level. The topics, which are listed in Table 1.1, span a wide range of levels from the introductory to the advanced. While the primary focus of this book is to demonstrate how the computer can enhance the study of physics, it is also possible to use the set of physics topics to illustrate various mathematical and/or computer methods. Thus, the topics in Table 1.1 have also been listed according to category of computer method.

We shall present these topics in Chapters 2-5. In Chapter 1 we consider the role of the computer in physics, the basic principles of computers and computer programming, and a sample problem. In sections 1.4-1.6 we introduce one of the languages used in communicating with the computer.

1.1 Computer Applications in Physics

The computer has long been recognized as an important tool for use in physics research. More recently it has found increasing use as an aid in physics teaching. This trend is almost certain to increase, as lower computer costs bring more powerful computing facilities within the reach of most institutions and as physicists continue to apply their research expertise in their teaching. While a sharp distinction cannot be made between computer applications in research and in teaching, generally research applications require more sophisticated hardware (i.e., computers and related equipment) and software (i.e., computer programs and documentation).

Computer Uses in Research

The computer has its greatest range of uses in the area of experimental physics. Starting with the initial planning phase, the computer is sometimes used to determine the feasibility of an experiment given the constraints imposed by the laws of physics, the precision of available apparatus, and monetary or other resources. This determination can be made by simulating the experiment using a computer model. To optimize an experimental design, it is necessary to determine values for a number of experimental parameters. (For example, in a scattering experiment we must choose the number, size, type, and placement of

Table 1.1

Physics Topics	**Associated Mathematical and/or Computer Methods**
1. Equipotential Surfaces for Two Point Charges	Computer Graphics and Intensity Scaling
2. Potential for a Charged Thin Wire	Numerical Integration by Trapezoidal and Simpson's Rules and Interpolating Polynomials
3. Discharge of a Capacitor in an RC Circuit	Numerical Solution of Differential Equations Using Euler's Method
4. The RLC Series Circuit	Numerical Solution of Differential Equations Using *Improved* Euler's Method
5. Superposition of Waves	Fourier Analysis and Synthesis of Wave Forms
6. Waves in Two Dimensions: Doppler Effect, Shock Waves, and Interference	Computer Graphics and Superposition of Waves
7. Solution of the Schrödinger Equation	Use of Boundary Conditions to Select Proper Solutions: Eigenfunctions and Eigenvalues
8. Random Processes	Generation of Random Numbers by Computer: the Power Residue Method
9. Center of Mass Coordinates	Numerical Integration Using the Monte Carlo Method
10. Fission Chain Reaction: the Critical Mass	Simulation of a Random Process Using the Monte Carlo Method
11. The Approach to Equilibrium: the Second Law of Thermodynamics	Simulation of a Random Process Using the Monte Carlo Method
12. Fitting of Experimental Data	Least Squares Fitting
13. Measurement of Bubble Chamber Photographs	"Missing Mass" Calculation

the particle detectors.) In the computer simulation, one parameter is varied at a time in order to determine its effect on the experimental results. This technique often makes it possible to determine the best possible experimental arrangement and the accuracy of the experiment *before* it is actually performed.

For many experiments, particularly those in nuclear and high energy physics, the computer can collect the data: signals produced by the apparatus are fed directly into the computer. In experiments involving large quantities of data coming in at a high rate, the computer can be used to make certain tests as the data come in, possibly rejecting data that fail to meet certain criteria, possibly looking for specific patterns in the data. This subject of "pattern recognition" by computer is of great interest to researchers in many scientific fields because of its wide applicability.

If the data from an experiment are examined by computer prior to being recorded, it becomes possible to use the computer to continually monitor the status of the equipment. For example, when an abnormal condition is detected, the computer can alert the experimenter by typing out a message. In sophisticated systems the detection process is extended, and the computer automatically corrects the abnormal condition it has detected.

There are many other computer uses in experimental physics including book-keeping—storing all the data from an experiment or compiling results from a number of experiments; fitting and hypothesis testing—determining if the data can be unambiguously categorized; generating graphs or other displays using the data; modeling—determining if the data are consistent with various theoretical models.

In theoretical physics, the range of computer uses is not quite as extensive. Computers can be used to perform various kinds of calculations, for example, evaluating definite integrals and solving differential equations. However, most theoretical physicists formulate theories and do the necessary algebraic manipulation of equations without the aid of the computer. The computer has so far proved most useful in generating numerical results using a previously derived theoretical model.

Computer Uses and Misuses in Physics Teaching

One way to distinguish the various computer applications in teaching is by the amount of detail of computer operation the student has to know. The two extremes can be illustrated by the problem-solving mode, in which a student writes his own computer programs to solve physical problems, and the simulation mode, in which a student inputs values of parameters and obtains a result, using a program that he has no knowledge about. The applications in this book are intended to cover a broad range of the spectrum, as they are suitable for use by both the student who knows little or no computer programming, as well as the experienced programmer. The beginner can use the programs in the book simply as "canned" programs. For the student who has (or wishes to gain) experience in writing computer programs, a number of modifications are suggested for each of the programs.

Many of the computer uses in physics teaching parallel the research uses. Computers can perform calculations, analyze data from experiments, perform simulations, and produce graphs and other visual material (even movies). Most physicists would agree that some uses of the computer in teaching physics are sounder than others, and that some uses should be avoided. The use of the computer in teaching physics either enhances the subject matter and stimulates interest, or it has the reverse effect—*depending on how the computer is used.* One example of poor use of the computer is to have it completely analyze the data from an experiment. If a student simply watches the computer collect and analyze the data and then type out the result, it is not clear what, if any, physics he has learned. Worse yet, he may get the impression that there is no point to understanding the analysis of data if he can always get the computer to do it. In general, the computer should never be used to avoid thinking through a problem. Of course, in an experiment involving a tedious analysis of a large amount of data, it would be very desirable to have the computer analyze the data *after* the student understands the steps in the process (i.e., after he does some of the analysis by hand).

1.2 Computers and Computer Programming

The computer is in one sense a very limited device, despite its varied and sophisticated applications. The basic operations which a computer can perform include little more than the simple arithmetic operations (addition, subtraction, multiplication, and division) and logical comparisons. Sophisticated applications are possible only because many complex problems can, after careful analysis, be reduced to a large number of such elementary operations.

Components of a Computer

A simplified block diagram of the components of a computer is shown schematically in Figure 1.1. The Central Processing Unit (CPU) is the component that controls the elementary arithmetic and logical operations. The CPU is also the control unit for all the other computer components with which it communicates in four ways:

1. Commanding various input devices to send information to the CPU (Examples of input devices include card reader, typewriter, magnetic tape unit.)

2. Sending information to various output devices (Examples include card punch, typewriter, printer, magnetic tape unit, and even an oscilloscope or plotter for drawing pictures.)

3. Sending information to the memory unit for storage

4. Taking previously stored information out of the memory unit

Figure 1.1   Components of a computer

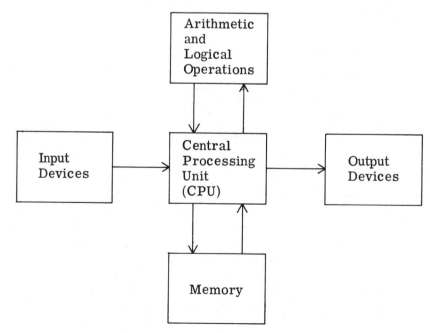

The unit of information processed by the computer is known as a *word*. A word can be either a number, a sequence of letters, or some other special symbols such as $! ? *. The computer memory is simply an *internal* medium where coded information can be stored. (Information can be stored externally on such media as punched cards, magnetic tape, or the printed page.) Most computer memories store information magnetically, using small magnetic "cores" or "thin films." As in the case of information stored externally on a magnetic tape, the information can be stored, retrieved, or erased from the memory. Each word (the unit of information) is stored in a particular location within the memory designated by a *memory address*. The maximum number of words that can be stored in the computer memory varies with different machines. There is also a specific limit on the number of digits or symbols that can be stored in each word; this limit is also machine-dependent. On many computers, words are represented in binary (base two) notation, with the value of each of the binary digits (bits) designated by the state of a magnetic core which can have one of two possible orientations.

Computer Programs

All the operations which a computer performs are specified by a computer program. A *program* is a coded list of instructions which is fed into the computer on punched cards or some other medium. Usually, instead of carrying out or *executing* each separate instruction as it comes in, the CPU temporarily stores all the instructions in memory. At a later time it retrieves the instructions from memory one at a time and executes them. In order to create a computer program one normally follows a number of steps:

1. Define the Problem.   State in the clearest possible terms the problem you wish to solve. It is impossible to write a computer program to solve a problem which has been ambiguously or imprecisely stated.

2. Devise an Algorithm.   An *algorithm* is a step-by-step procedure for solving the problem. Each of the steps must be a simple operation which the computer is capable of doing. A universally-used way to represent an algorithm is the *flow chart* or *flow diagram,* in which the steps in the procedure are shown as a number of interconnected boxes with arrows indicating the sequence of steps.

3. Code the Program.   The steps in a flow chart can be translated into a series of instructions to the computer which comprise a computer program. There are many languages in which computer programs can be coded (i.e., written, each with its own syntax, vocabulary, and special features). One of the most widely used languages for scientific applications is FORTRAN (formula translation), which is discussed later in this chapter. Other languages, such as BASIC, APL, and PL/1 are also available on many computer systems.

4. Debug the Program.   Most programs of any length don't work properly the first time they are run and must be *debugged*. Often during the debugging phase, errors and ambiguities in the original statement of the problem reveal themselves, calling for basic revisions in the solution algorithm. We shall return to the subject of debugging in more detail after discussing the FORTRAN language.

From the preceding discussion of the steps necessary to create a computer program, it should be clear that in order to write the program we must actually know how to solve the problem without a computer! That is, once we devise an algorithm, we could in principle perform each operation in sequence and arrive at the solution. However, in practice many algorithms have so many steps that only a computer, which can do perhaps millions of arithmetic operations per second, can perform all the operations without error and in a reasonable amount of time.

Flow Diagrams

Flow diagrams can prove useful in solving a wide variety of problems even when we do *not* wish to create a computer program. For example, we can arrange a set of driving instructions to reach a particular destination in flow diagram form as shown in Figure 1.2. From this example it should be clear that the steps indicated in each of the boxes in the flow diagram must be carried out in the specified order. Should the order of any steps be changed, a person following these instructions would most likely not reach the desired destination.

Most flow diagrams are more complex than the "straight line" diagram of Figure 1.2. Usually, flow diagrams have one or more *branches*, points at which one of a number of paths is taken, depending on some condition. In the most common form of a two-way branch (see Figure 1.3a), the "yes" path is taken if the answer to the question (contained in the oval) is affirmative, and the "no" path is taken if the answer is negative. Suppose, for example, we wish to compute the income tax a person would pay under the simplified assumption that the tax is computed as 20% of income in excess of $3000. The formula for computing a person's tax:

$$TAX = 0.2 \, (INCOME - 3000)$$

applies only if the income is not less than $3000, since it would otherwise yield a negative tax. Hence, we require a two-way branch to test if the person's income exceeds $3000. Two alternative methods of specifying such a branch are shown in Figures 1.3b and 1.3c. In Figure 1.3c, which we refer to as an *arithmetic branch*, one of two paths is taken depending on whether the expression in the oval is positive (> 0) or not (≤ 0). Note that arithmetic branches can, if desired, be readily extended to allow for a three-way branch ($> 0, = 0, < 0$). The entire flow diagram for computing a person's income tax using the arithmetic branch is shown in Figure 1.4. The backward arrow symbol ($\leftarrow$), used in several places in Figure 1.4, indicates that a quantity (TAX) is to be assigned a particular numerical value. It might seem more natural to use the equal sign in place of the backward arrow, but we shall see later why the backward arrow is a more appropriate symbol in this context.

Using different shapes to denote particular operations in a flow diagram such as that in Figure 1.4 is a conventional technique. The meaning of the various shapes is given in Figure 1.5. The circled "begin" and "end" have an obvious meaning. The box shaped like a punch card (with one corner cut off) means that a card which contains numerical values for certain specified quantities is to be read by the computer. This is one convenient method of assigning numerical values to quantities without explicitly specifying these values in the flow diagram. The card shape may also be used as a more general symbol for *input*, even if some device other than a card reader is used.

Figure 1.2   An example of a straight line flow chart

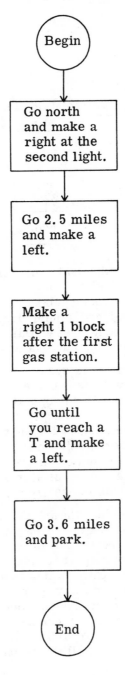

Figure 1.3   *Examples of branches*

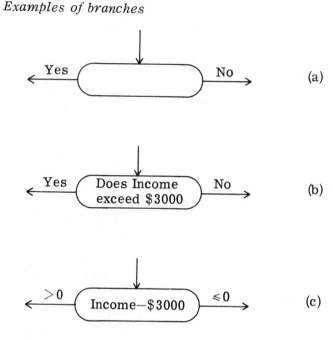

As noted previously, the oval represents a branch, and the paths leading from the oval are labeled either yes/no or $>0/<0/=0$, depending on which type of branch is used. The rectangular "assignment box" is used to indicate that a quantity is to be assigned a numerical value. If the quantity is calculated from an algebraic formula, it is essential that all quantities in the formula have been previously assigned numerical values.

The *output* box is drawn to symbolize a piece of paper hastily torn off the printer, but it may also be considered a general symbol for output. The small circle does not designate any operation in a flow diagram, but is used to indicate a reference point and usually encloses some arbitrary number. For example, in a very complicated flow chart, if it is necessary to use more than one page, then numbered reference points can show places where the flow chart leaves one page and resumes on the next. Another use of numbered reference points is in loops. A *loop* is used in a flow diagram when it is desired to repeat a series of steps. This is usually represented by a hexagonal "iteration box." *

In Figure 1.6 we see how the iteration box is used to indicate a loop: the contents of the iteration box specifies that all the steps up to the circled reference point ⑩ are to be repeated 25 times. The dotted line indicates the return path which closes the loop. Apart from the (dotted) return path which closes a loop, there can be only one path leading *away from* any circled reference point. The arrow leading downward away from the circled reference point indicates the path to be taken after the loop has been "satisfied" (cycled 25 times in this case). As an example of how a loop can be used in a flow diagram, suppose we wish to compute the income tax for 100 taxpayers. As shown in Figure 1.7, we can achieve this by enclosing all the steps of flow diagram 1.4 inside a loop.

---

* A hexagon is also used to indicate a *subroutine* or sub-algorithm contained within or as part of another algorithm.

Figure 1.4   An example of the use of a branch

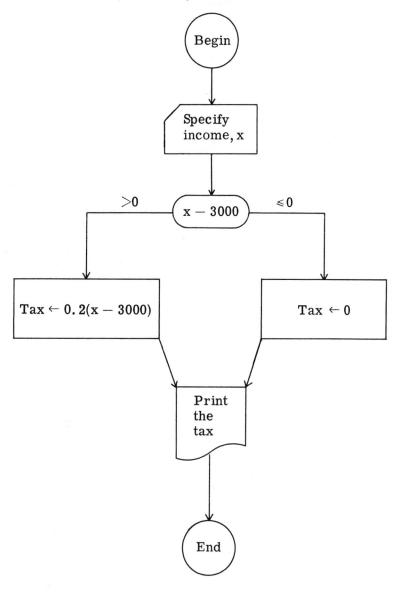

Figure 1.5   Shapes used in flow diagrams

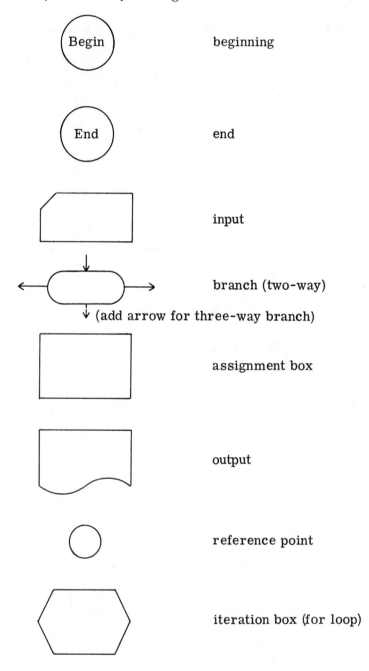

Figure 1.6   Use of an iteration box to define a loop

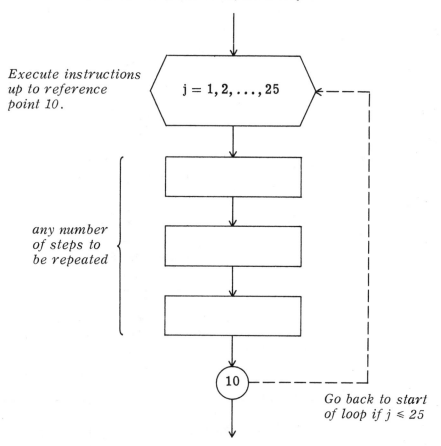

As a final example, we show in Figure 1. 8 the prototype for many flow diagrams used to perform a calculation. The flow diagrams for all the problems listed in Table 1. 1 are based on this prototype, which includes the following steps:

1. Read a data card containing numerical values for a number of input parameters.

2. Print the input parameters (as a check).

3. Test the input parameters to see if they are in their allowed range, and stop if this is not the case.

4. Do some calculation using this set of input parameters.

5. Write out the results.

6. Read the next data card if any remain to be read, otherwise stop.

The above sequence causes the calculation to be repeated for a number of sets of input parameters, terminating when a data card containing invalid parameters (possibly a blank card) is read, or until no more data cards remain.

In the next section, we shall consider other examples of flow diagrams in the context of a particular physics problem.

Figure 1.7   An example of the use of a loop and a branch

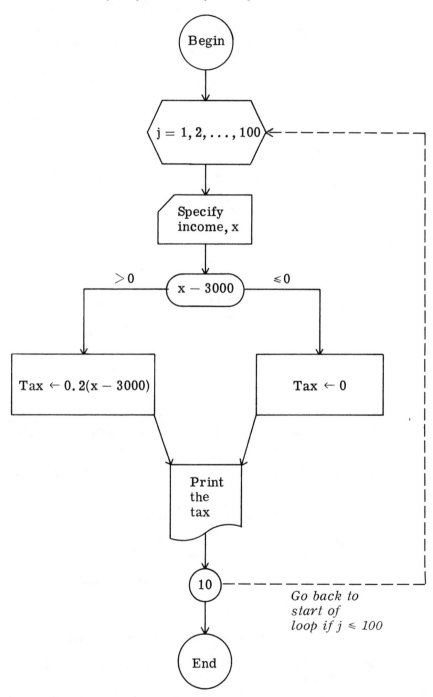

Figure 1.8   Prototype for many flow diagrams

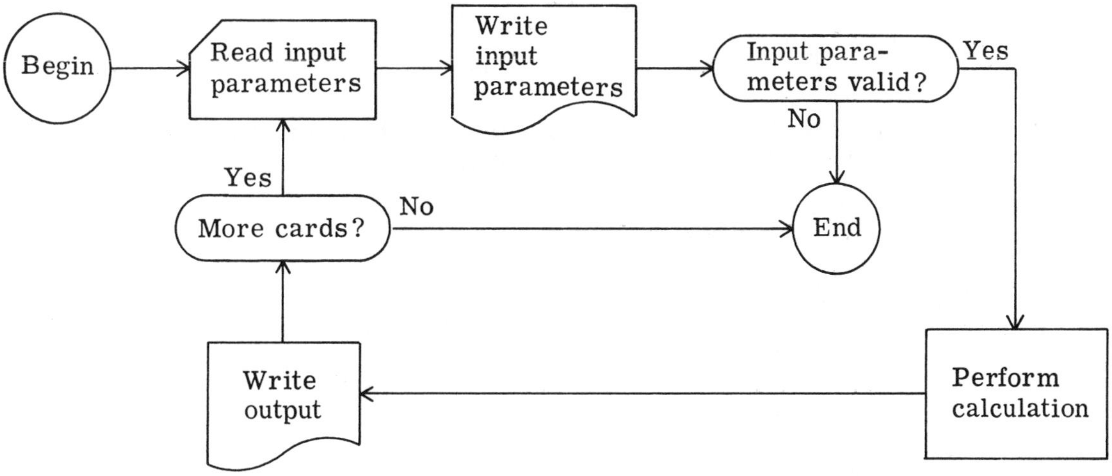

```
Begin → Read input → Write input → Input para-  Yes →
         parameters    parameters    meters valid?
            ↑                            No
           Yes                            ↓
        More cards? ── No ──────────→    End
            ↑
        Write output ←─────────────── Perform calculation
                                            ↑
```

1.3 A Sample Problem

To illustrate the capabilities (and limitations) of the computer in solving physics problems, let us consider the following electrostatics problem: find the location of the point on a line joining two point charges q_1 and q_2, separated by a distance of 1 meter, where the resultant electric field is zero.

Assuming that the charges are of like sign, the point at which the field vanishes must lie in the one-meter interval between the charges. As can be seen in Figure 1.9, the magnitude of the net electric field at the point P can be written

$$E = E_1 - E_2 = \frac{q_1}{r_1^2} - \frac{q_2}{r_2^2}.$$

(1.1)

Figure 1.9   The electric field due to two positive charges

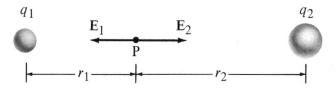

(We have omitted the Coulomb force constant $k = \dfrac{1}{4\pi\epsilon_0}$ which normally appears in the rationalized mks system of units, since it will in no way affect our result.) To find the point at which the net electric field vanishes, let us express the net

field as a function of one variable. We let $r_1 = x$, so that $r_2 = 1 - x$. Thus, the net electric field can be written

$$E(x) = \frac{q_1}{x^2} - \frac{q_2}{(1 - x)^2} .$$ (1.2)

Our original problem has thus been transformed into a purely mathematical one: find the value of x which satisfies the equation

$$E(x) = 0.$$ (1.3)

We have

$$\frac{q_1}{x^2} - \frac{q_2}{(1 - x)^2} = 0,$$

and therefore

$$(1 - x) = \left(\frac{q_2}{q_1}\right)^{1/2} x.$$

We solve for x:

$$x = \frac{1}{1 + \left(\dfrac{q_2}{q_1}\right)^{1/2}} .$$ (1.4)

Having described the procedure for finding a solution by hand, let us now consider how we could devise an algorithm suitable for a computer solution. It is possible, though fairly difficult, to write a program to do the algebraic manipulation necessary to solve an equation such as 1.3 symbolically. Usually, we use the computer to find a numerical solution to an equation rather than a symbolic one.

An Algorithm

In the simplest type of algorithm we could make use of the algebraic solution (equation 1.4) in order to determine a numerical value for x using specified numerical values of q_1 and q_2. The three steps in the algorithm are shown in the flow chart in Figure 1.10.

Often we want to calculate a specific quantity which depends on a number of parameters. We can see how the result depends on a parameter by choosing a range of numerical values for each one. In the present case, for example, we might wish to have the computer calculate values for x using many sets of values for q_1 and q_2. This can be accomplished by putting the three steps of the algorithm inside a loop, as shown in Figure 1.11. The exit from the loop is provided by a branch, at which a test is made for the condition $q_1 = 0$. By putting an extra data card on which q_1 has the value zero after the last data card, the computer will stop after reading this last data card. Note that in order to save writing we have used the shape of each box in Figure 1.11 to indicate the nature of each operation, making it possible to leave out the words.

Figure 1.10   *Flow chart to compute x*

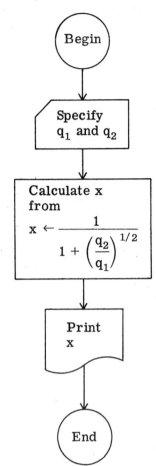

In devising an algorithm and writing a computer program based on the algorithm, it is necessary to consider all possibilities. Otherwise, a procedure may yield nonsensical results under unanticipated conditions. For example, the procedure specified by the flow chart in Figure 1.11 gives a correct result only if q_1 and q_2 have the same sign. If q_1 and q_2 have opposite signs, equation 1.4 gives an absurd result, indicating that the procedure must be modified to handle this case.

If the two charges have opposite signs, the point at which the net electric field is zero lies outside the one-meter interval between the charges, as shown in Figure 1.12. In this case, the magnitude of the net electric field can be written

$$E = E_1 + E_2 = \frac{q_1}{r_1^2} + \frac{q_2}{r_2^2}.$$

To express the net field as a function of a single variable, we use $r_1 = x$ and $r_2 = 1 + x$ to obtain

$$E(x) = \frac{q_1}{x^2} + \frac{q_2}{(1 + x)^2}. \tag{1.5}$$

Figure 1.11   Flow chart to compute x for many values q_1 and q_2 ($q_1 = 0$ signals end of computation)

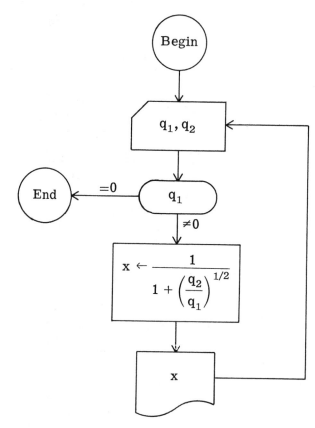

As before, the problem is to determine the value of x, for which $E(x) = 0$. With the aid of a little algebra, we find

$$x = \frac{1}{1 - \left(-\dfrac{q_2}{q_1}\right)^{1/2}} \, . \tag{1.6}$$

Figure 1.12   The electric field due to two charges of opposite sign

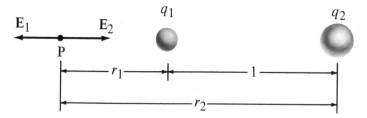

Let us use this result to modify the algorithm for calculating x, so that it works regardless of the signs of the two charges. If q_1 and q_2 have the same sign, we wish to use equation 1.4; if q_1 and q_2 have opposite signs, we wish to use equation 1.6. We accomplish this by including a two-way branch in the flow chart, which tests whether the ratio q_2/q_1 is a positive number. The modified algorithm is shown in Figure 1.13.

Figure 1.13   Flow chart to compute x for arbitrary signs of q_1 and q_2

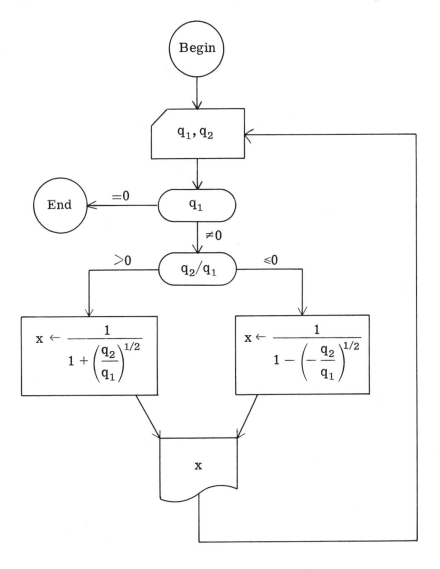

The role played by the computer in solving this problem is that of a high speed calculator which obtains numerical values for x using one of two specified formulas according to the signs of the charges. Other algorithms can be devised to solve this problem which do not require that formulas for x be specified, but only the form of the equation $E(x) = 0$. While it is somewhat difficult to obtain an algebraic solution of an equation by computer, there are many methods for finding numerical solutions, one of which is discussed in the next section.

An Alternative Algorithm

If we place the charges on the horizontal axis in a coordinate system, we find the point where the net electric field vanishes by finding where the function E(x) (given by either equation 1.2 or 1.5) crosses the x-axis. For two charges of like sign located at x = 0 and x = 1, this point lies between these values, as illustrated in Figure 1.14, which is a plot of equation 1.2 for particular positive values of q_1 and q_2. There is a very simple "search" procedure which we can use to find the crossing point x_0 to any desired accuracy. (As described, the procedure applies to the case of two positive charges, and with only a slight modification, to the case of two negative charges as well.)

As can be seen in Figure 1.14, if we compute the value of the function at some particular value of x and we find that E(x) is positive, then the crossing point

Figure 1.14   E(x) against x for positive q_1 and q_2

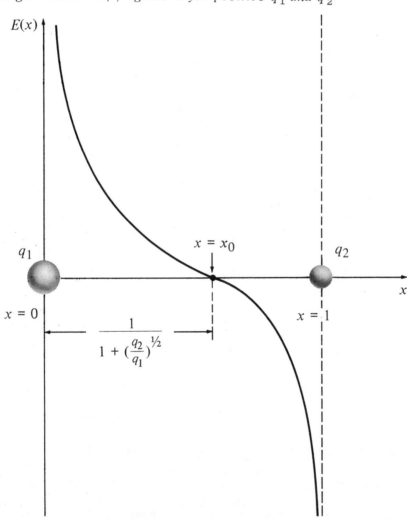

Figure 1.15   Decision tree algorithm for solving   E(x) = 0

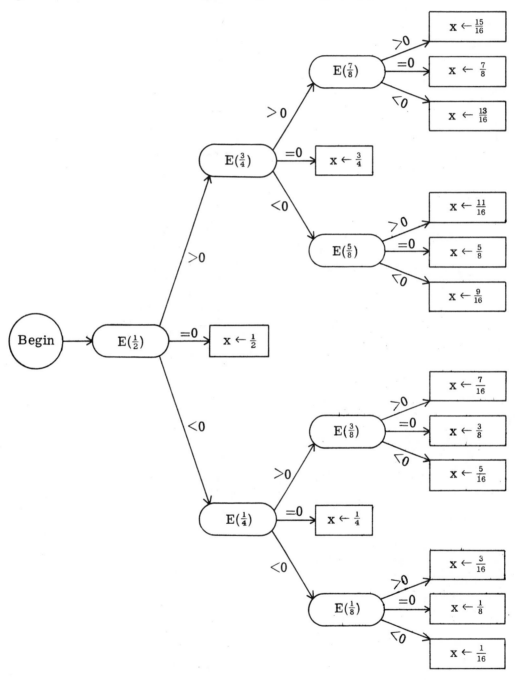

Figure 1.16   Decision tree algorithm in compact form

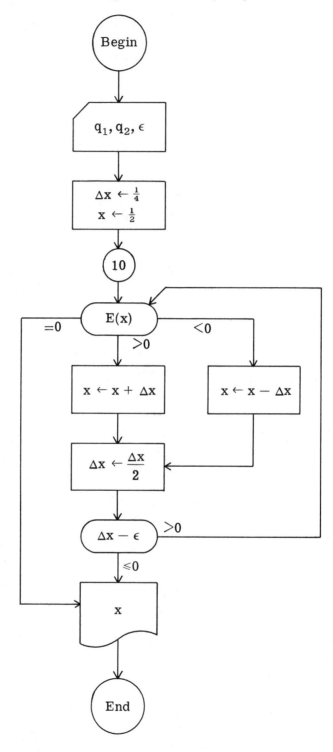

x_0 must be to the right of this x-value, i.e., $x_0 > x$. Similarly, if we find that $E(x) < 0$, then the crossing point must be to the left of this value, i.e., $x_0 < x$. (Should it happen that $E(x) = 0$, then $x_0 = x$.) Thus, if we determine the sign of the function $E(x)$ at a series of points, we can close in an interval which is known to contain the crossing point x_0. Clearly, the most efficient search procedure is to evaluate $E(x)$ at the *midpoint* of the previous interval which is known to contain x_0. This permits us to cut the size of the interval known to contain x_0 in half after each evaluation. This algorithm, depicted in Figure 1.15, is an example of a *decision tree*. After three levels of the decision tree, the value of x_0 is known to lie on an interval no more than $\frac{1}{8}$ meter in length.* For example, if the branches in the decision tree should bring us to the box labeled $x \leftarrow \frac{9}{16}$, we could then assert that $x_0 = \frac{9}{16} \pm \frac{1}{16}$.

Since the number of branches in the decision tree doubles at each level, showing the flow diagram for very many levels is cumbersome. Fortunately, there is a way to represent the decision tree algorithm using a much more compact flow chart. The simplification makes use of the relation between the x-values appearing in each branch of the flow diagram in Figure 1.15. To find the x-value appearing in any branch, we take the x-value from the previous branch and either add Δx or subtract Δx, depending on whether the greater-than-zero or less-than-zero branch was taken. (The value of Δx is taken to be $\frac{1}{4}$ initially, and it is reduced by a factor of two each time through the loop.)

This rule has been incorporated in the flow chart of Figure 1.16 in which a single branch replaces all the branches in the decision tree. The branch appears inside a loop, and the x-value which is used in the branch is modified each time through the loop in a manner that depends on the results of the previous branch in exactly the same way that the contents of each branch change in going from one level of the decision tree to the next. Note that each time the loop is executed, the value of x gets closer to the crossing point x_0, which is the solution to the equation $E(x) = 0$. In order to exit from the loop, the computer must find $\Delta x \leq \epsilon$, in which case the value of x differs from x_0 by no more than 2ϵ. We can therefore determine x_0 to any desired accuracy by specifying a sufficiently small value for the error parameter ϵ. Of course, the smaller the value we choose for ϵ, the more times the loop must be executed to obtain the desired accuracy.

The algorithm depicted in Figure 1.16 should be studied carefully, since it illustrates an important extension in the use of a loop—modifying the instructions in the loop each time through the loop.

Comparison of the Two Algorithms

Let us consider the relative merits of the two algorithms represented by the flow charts of Figures 1.10 and 1.16. The first algorithm leads to an exact numerical solution to the equation $E(x) = 0$, but necessitates solving the equation by hand algebraically, in order to obtain formulas for x. The second algorithm produces a numerical solution which is only approximate, but does not necessitate first solving the equation algebraically. An approximate result from the second algorithm is not a real drawback, since any desired accuracy can be obtained by choosing a sufficiently small value for the parameter ϵ. For example, if we choose $\epsilon = 10^{-8}$, then the computed result will be correct to this

---

* If $E(x)$ should happen to be zero, then x_0 is exactly determined.

Figure 1.17   *Generalized root-finding algorithm for any function E(x)*

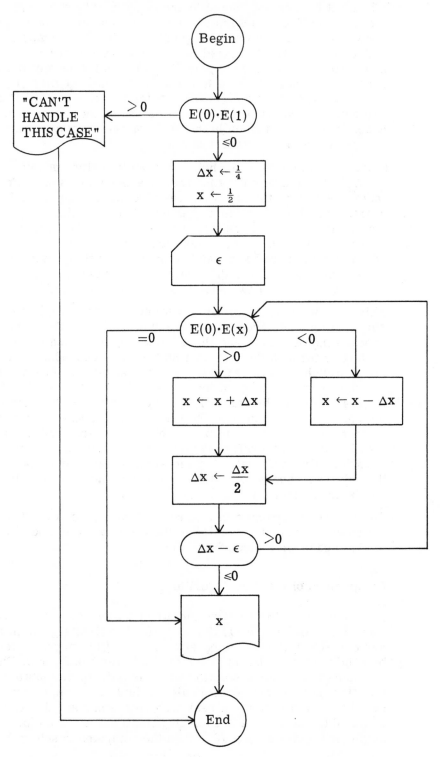

amount (i.e., eight significant figures if $x \geq 0.1$). There is, however, a maximum possible accuracy on any computer due to the finite word size.* The "exact" result obtained using the first algorithm is subject to this same limitation.

Furthermore, the second algorithm has one important advantage over the first: It does not require the equation $E(x) = 0$ to be first solved algebraically. This may seem like a trivial advantage, since only a few algebraic steps are needed to solve the equation. However, the second algorithm can be used for any function $E(x)$, which changes sign between $x = 0$ and $x = 1$, even if no algebraic solution to the equation $E(x) = 0$ exists.

In its present form, the algorithm assumes the root to lie at a larger x-value $(x + \Delta x)$ if $E(x)$ is positive, and a smaller x-value $(x - \Delta x)$ if $E(x)$ is negative. The basis of this assumption is the prior assumption that q_1 and q_2 are both positive. We can easily generalize the algorithm to avoid this restriction. The generalized root-finding algorithm shown in Figure 1.17 can be used to locate a root of any function $E(x)$ for which $E(x = 0)$ and $E(x = 1)$ are finite and of opposite sign, i.e., for which the function has an *odd* number of roots between $x = 0$ and $x = 1$.

Since the algorithm works only if the function $E(x)$ has opposite signs at $x = 0$ and $x = 1$, we immediately quit if this is not the case, i.e., if the product $E(0) \cdot E(1)$ is positive. The remainder of the flow diagram is the same as that in Figure 1.16, except that at the branch we now test the value of the product $E(0) \cdot E(x)$ instead of $E(x)$. This has the desired effect of reversing the >0 and <0 branches, in the event that $E(0)$ is negative and $E(1)$ is positive.

Another extension of the decision tree algorithm is to use it for locating the point at which a function $E(x)$ has its extreme value, a maximum or a minimum, in the interval $0 < x < 1$. At a maximum or minimum, the derivative $E'(x) = \frac{d}{dx} E(x)$ vanishes. Thus, we can locate an extreme value by finding the root of the function $E'(x)$, using the algorithm in Figure 1.17.

A Hill-Climbing Algorithm

The same decision tree algorithm is the basis for a "hill-climbing" method, which can be used to locate the point at which a function of two variables has a maximum or a minimum. Suppose the function $F(x, y)$ has a maximum or a minimum somewhere inside the square region defined by $0 < x < 1$, $0 < y < 1$. The function is represented in Figure 1.18 by a series of contours, each of which corresponds to the locus of all points for which $F(x, y) = $ a constant. At a maximum or a minimum, we have $F_x(x, y) = 0$ and $F_y(x, y) = 0$, where the functions $F_x(x, y)$ and $F_y(x, y)$ are the partial derivatives of $F(x, y)$ with respect to x and y. If we start at the center of the square, $(x = \frac{1}{2}, y = \frac{1}{2})$, we can try to approach the maximum or minimum by changing x to $x + \Delta x$ if $F_x(0, y) \cdot F_x(x, y)$ is positive and to $x - \Delta x$ if $F_x(0, y) \cdot F_x(x, y)$ is negative. We then follow the

---

* On most computers the number of significant figures accuracy can be increased by using two or more memory words for each variable—see section 1.6.

same procedure for the y-coordinate, changing y to y + Δy if $F_y(x, 0) \cdot F_y(x, y)$ is positive and to y — Δy if $F_y(x, 0) \cdot F_y(x, y)$ is negative. The step sizes Δx and Δy are both initially $\frac{1}{4}$, and are reduced by a factor of two after each step. An example of the zigzag path to the summit (or valley) which might occur when this procedure is followed is illustrated in Figure 1.18.

Figure 1.18   A hill-climbing algorithm

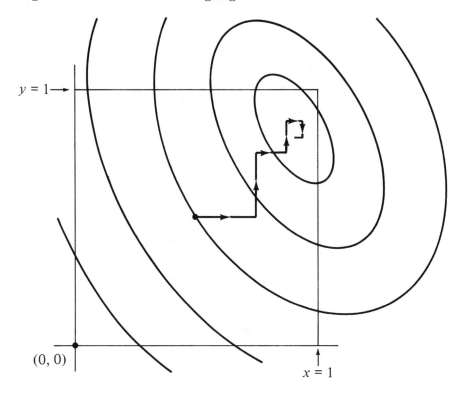

1.4 Introduction to FORTRAN IV

FORTRAN is actually not a single language but a collection of closely related computer languages. Moreover, each of these languages has a number of "dialects" with minor variations from one type of computer to another. The core of each language is common to a wide range of computers, and a number of extra features can be implemented on the more powerful computers. All the programs in this book are written in the FORTRAN IV language and have been tested on an IBM 1130 having 8000 words of memory. This relatively modest memory size restricts the language to its essential core. As a result, none of the programs use features of FORTRAN IV that are not common to almost all computers that use the language.

The aim of this section is to cover enough of the fundamental FORTRAN instructions for you to understand FORTRAN programs and write your own as soon as possible. A working knowledge of the language will of necessity not touch on some of the finer points which can be found in many textbooks on FORTRAN IV. However, probably the most ineffective way to learn FORTRAN IV, as in the case of any foreign language, is to try using it only after learning all the rules. A much better approach is to use the language as you learn it. In view of this, instead of simply enumerating the rules of the language, we shall consider several examples of simple FORTRAN IV programs and discuss the rules as the need arises.

The individual instructions in a FORTRAN program are known as FORTRAN *statements*. In many cases there is nearly a one-to-one correspondence between the boxes of a flow chart and the statements in a program based on the flow chart. In the next section we shall see how previously discussed flow charts can be translated into computer programs.

A Sample FORTRAN IV Program

The three main statements* comprising a computer program based on the flow chart of Figure 1.10 are essentially the following:

```
READ Q1, Q2
X = 1.0/(1.0 + (Q2/Q1)**0.5)
WRITE X
```

The first statement is an instruction to the computer to read in numerical values for the parameters q_1 and q_2. (The use of Q1 for q_1 and Q2 for q_2 is necessary since lower-case letters and subscripts are not permitted in FORTRAN.) The second statement is an instruction to compute a value for x, according to the formula

$$x = \frac{1}{\left(1 + \left(\dfrac{q_2}{q_1}\right)^{1/2}\right)}.$$

The translation of this formula into a FORTRAN statement is accomplished by simply changing the symbols for each variable, and using the FORTRAN operators: "/" to indicate division, "+" to indicate addition, and "**" to indicate exponentiation, i.e., raising to a power. The use of 1.0 instead of 1 and 0.5 instead of $\frac{1}{2}$ is necessary for reasons to be explained later. The third statement is an instruction to print the value of X calculated in the previous statement. The form of the READ and WRITE statements are not quite correct as they appear and this will also be discussed later.

---

* Each of which is keypunched onto a separate card; the cards are fed sequentially into the computer.

Arithmetic Operations in FORTRAN

Three of the five possible arithmetic operations in FORTRAN have already been mentioned. We list the entire set below:

Operation	FORTRAN symbol
exponentiation	**
multiplication	*
division	/
addition	+
subtraction	—

When two or more operations appear in a FORTRAN statement, parentheses should be used to indicate the order in which the operations are to be performed. For example, in FORTRAN,

$$C = \frac{A + B}{2}$$

is written

$$C = (A + B)/2.0.$$

If the parentheses are left out, we have

$$C = A + B/2.0$$

which the computer would interpret as

$$C = A + \tfrac{1}{2}B.$$

As a general rule, if parentheses are not used to signify the order in which operations are to be performed, the following order of priority is assumed:

1. exponentiation
2. multiplication or division
3. addition or subtraction

In the case of two operations of equal priority, they are performed in a *left to right* order. For example,

$$A**0.5 + B/C*D$$

is interpreted as

$$A^{1/2} + \frac{BD}{C}.$$

Use of the Equal Sign in FORTRAN Statements

Any FORTRAN statement containing an equal sign is known as an *assignment statement*. The general form of the assignment statement is

variable = expression,

where "expression" stands for any mathematical expression involving variables and constants or possibly just a single variable or constant. Note that to the left of the equal sign there must be a *single* variable. Thus, the following are valid assignment statements:

    X = 1.0
    C = A + B
    Z = (A + 1.0)**2 + C*X

while the statements obtained by interchanging the left and right sides of the three equalities:

    1.0 = X
    A + B = C
    (A + 1.0)**2 + C*X = Z

are all invalid. A single variable must appear to the left of the equal sign because the assignment statement is an instruction to the computer to

1.   find a numerical value for the expression to the right of the equal sign, and then

2.   store that value in the memory location reserved for the variable which appears to the left of the equal sign.

Note that each variable appearing in a FORTRAN program is associated with a different location in the computer memory. For example, the statement

    C = A + B

is an instruction to add the numerical values stored in memory locations reserved for variables A and B, and then store the result in a memory location reserved for the variable C. Clearly, for the statement to be meaningful, it must come after other statements which instruct the computer to store numerical values in the memory locations reserved for variables A and B. One way to have A and B assigned numerical values is to use a READ statement:

    READ A, B

which is an instruction to the computer to read numerical values for A and B, possibly from a data card. Alternatively, A and B might be assigned numerical values in several previous assignment statements. For example,

    A = 2.0
    B = A + 1.0
    C = A + B

After these three statements are executed, the memory locations corresponding to the variables A, B, and C contain the numerical values 2.0, 3.0, and 5.0, respectively.

If a variable appears to the left of an equal sign in more than one assignment statement, its value is modified as each statement is executed. For example, after the computer executes the two statements

    X = 1.0
    X = X + 1.0

X has the value 2.0, until the execution of any other assignment statement in which X appears to the left of an equal sign (or a statement such as READ X). Note that the statement   X = X + 1.0   is a perfectly valid assignment statement which means "Add 1.0 to the present value of X." The use of the equal sign in FORTRAN is quite different from its use in mathematics, since as an equation, x = x + 1 is meaningless. The real meaning of the assignment statement might have been better conveyed if the symbol "←" were used in place of the equal sign:

$$X \leftarrow X + 1.0$$

In any case, the equal sign is an established part of the FORTRAN language.

From the examples already considered, it should be clear that the order in which a series of FORTRAN statements appear can be very important. As a further illustration, suppose the following three statements appear in a program:

```
X = 1.0
X = X + 2.0
X = 2.0*X
```

After the computer executes these statements, X has the value 6.0. However, if the second and third statements are interchanged:

```
X = 1.0
X = 2.0*X
X = X + 2.0
```

X has the value 4.0 upon execution of the three statements. The instructions of a program are normally executed sequentially, corresponding to the simple straight line flow chart, such as that in Figures 1.2 or 1.10. In the next three subsections we discuss a number of ways to modify this normal sequence.

Statement Numbers

Any FORTRAN statement may be given an identifying number. The need for statement numbers becomes clear if we wish to provide a mechanism for including branches or loops in a program. As an example let us consider the following program which is an (approximate) FORTRAN representation of the flow chart in Figure 1.11.

```
10       READ Q1, Q2
         IF(Q1)30, 20, 30
20       CALL EXIT
30       X = 1.0/(1.0 + (Q2/Q1)**0.5)
         WRITE X
         GO TO 10
         END
```

Only three of the statements need to be given numbers (10, 20, and 30), since only these statements are referred to by other statements in the program. Of the seven statements in this program, three have been previously discussed. We now discuss the remaining four: the IF statement, the GO TO statement, the CALL EXIT statement, and the END statement.

IF Statement

The IF statement is the FORTRAN equivalent of a branch. The general form of the IF statement is

IF (expression) N_1, N_2, N_3

where "expression" stands for any FORTRAN expression, and N_1, N_2, and N_3 are three statement numbers of other statements in the program. The IF statement is an instruction to the computer to

1. find a numerical value for the expression in parentheses, and then

2. go to the statement numbered N_1, N_2, or N_3 for the next instruction, depending on whether the numerical value is negative, zero, or positive, respectively.

The IF statement thus permits a three-way branch. If only a two-way branch is desired, two of the three statement numbers N_1, N_2, and N_3 must be set equal. For example, in the case of the program we are presently considering, a two-way branch is provided by the statement

IF (Q1) 30, 20, 30

which instructs the computer to branch to statement 30 for its next instruction if the value of Q1 is either negative or positive, and to branch to statement 20 if it is zero.

The type of IF statement that we have discussed is known as the arithmetic IF statement. Another very useful type of statement is the logical IF statement, which we shall not discuss because it cannot be used in all versions of the FORTRAN IV language.

GO TO Statement

The general form of the GO TO statement is

GO TO N

where N is the statement number of another statement in the program. The GO TO statement is an instruction, or command, to branch to the statement numbered N, unconditionally.

Thus the sixth statement in the program on page 28 commands the computer to go to statement 10 for its next instruction. This statement takes the place of the return arrow in the flow chart of Figure 1.11. Note that when a loop is closed using a GO TO statement rather than an IF statement, it is necessary that an IF statement appear somewhere within the loop, in order that the loop not be infinite. (In the present case, the statement

IF (Q1) 30, 20, 30

provides an exit from the loop when Q1 is zero.)

CALL EXIT, END, and CONTINUE Statements

The statement CALL EXIT is an instruction to stop the program since there is nothing further to be done. (On some computers the statement used for this purpose is STOP, and on still others it is END). The CALL EXIT statement, which can appear anywhere in the program, takes the place of the circled "End" appearing in the flow chart. Somewhat confusingly, the END statement is used to signify that there are no more statements, so that the END statement must always be the last one in a program. To make matters worse, some computers use the END statement to signify both an instruction to stop the program and an end to the list of instructions comprising the program. To indicate how the END statement can be used in these two ways, suppose we have a program for which we wish the program to stop somewhere in the middle, assuming some condition is encountered. On computers which have a CALL EXIT statement, we would have

.

.

.

        CALL EXIT
.

.

.

        END

On a computer which uses the END statement both ways this might become

.

.

.

        GO TO 40
.

.

.

    40   CONTINUE
        END

The CONTINUE statement, as the name implies, means "proceed to the next instruction," and is a *dummy statement*. Its only purpose is to provide a numbered statement to branch to from some other part of the program. A numbered END statement is *not* permitted. So it is not possible to replace the last two statements in the program with

    40   END

READ, WRITE, and FORMAT Statements

As noted previously, the READ statement is an instruction to read in values for some quantities from a data card or other input medium. The previous examples of READ statements have been deliberately oversimplified to temporarily avoid the tedious subject of FORMAT statements. To be complete, an instruction to read in numerical values for a series of quantities must specify the following:

1. the quantities to be read into the computer,

2. the input device to be used to read the data: card reader, magnetic tape unit, typewriter, etc.,

3. the detailed format of the data as it appears on the card or other input medium.

As an example of a complete READ statement, we have

READ (2, 100) Q1, Q2

The "2" appearing in parentheses indicates that the values to be read for the variables Q1 and Q2 are to be read from device number 2. (The assignment of particular numbers to each device is computer-dependent.) The second number in the parentheses after the comma (100) is the statement number of another statement in the program which specifies the detailed format of the data which is to be read. A FORMAT statement in the same program as the previous READ statement might be

100   FORMAT (2F10.5)

The "2F10.5" appearing in parentheses contains in coded form, the following information:

(2)   There are two quantities to be read.

(F)   They are both numbers with decimal points.

(10) Each number takes up ten spaces, including blanks, decimal point, and sign.

(.5) Each number has five digits after the decimal point.

This code is discussed in more detail on pp. 45-49. Unlike most other kinds of statements, it does not matter where in the program a FORMAT statement is placed. This is because the FORMAT statement is not actually an instruction to the computer to do anything (an *executable* statement), but rather a statement that is used as a reference by some READ or WRITE statement in the program. A good practice is to put all FORMAT statements in one part of the program for handy reference e.g., just before END.

The correct form of the WRITE statement is very similar to the READ statement. For example,

WRITE(3, 101) Q1, Q2

is an instruction to write out values for Q1 and Q2 on device number 3, according to the format specified by FORMAT statement number 101. Note that the same FORMAT statement could be referred to by more than one READ or WRITE statement if the format of the quantities being read or written is the same. As an example of the use of FORMAT statements in a program, we give below the complete version of the program that previously appeared on page 28 in a simplified form.

```
10      READ (2, 100) Q1, Q2
        IF(Q1) 30, 20, 30
20      CALL EXIT
30      X = 1.0/(1.0 + (Q2/Q1)**0.5)
        WRITE(3, 101) X
        GO TO 10
100     FORMAT(2F10.5)
101     FORMAT(F10.5)
        END
```

1.5 Running a Program on the Computer

In addition to the "stand-alone" computer on which one user at a time can run a program, there has been a trend in recent years toward large time-shared computer systems having a number of terminals (perhaps teletypewriters) that are connected to the computer directly or over long distance telephone lines. For many applications, it appears to a user at any one of the terminals that the entire computer is at his disposal, whereas the computer actually services each user in turn, devoting a fraction of its time to each one. These two types of computer systems (the time-shared multi-terminal system and the stand-alone computer) are best suited to two different modes of operation: *conversational* and *batch processing*.

Conversational and Batch Processing Modes

The time-shared system is best suited to a conversational type of computer usage. In this mode a user at a terminal first types in the appropriate "system commands" to the computer. In response, the computer generally causes messages to be typed back to the user. The nature of the messages will depend upon whether the computer can identify and comply with the user's commands. The system commands, like the statements of a FORTRAN program, are instructions to the computer. However, they are not part of the FORTRAN language and vary from one computer to another. Typical examples of system commands include instructions to load a specific program into memory, start execution of the program, sign-on or sign-off the terminal.

The stand-alone computer is best suited to the batch processing mode of use. In this type of operation, a user first punches the program on a deck of cards, one FORTRAN statement per card, using a key punch machine. Other media such as paper tape or magnetic tape may be used in place of cards on some computers, and our use of a card-oriented terminology is not meant to exclude these possibilities. The discussion which follows should be completely applicable in these cases with a minor change of words. For example, if a teletypewriter is used as the input device instead of a card reader, just replace "card" by "teletypewritten line," "card reader" by "teletypewriter," "card column" by "position within a line," etc. In the batch processing mode, a deck of cards, or perhaps a whole "batch" of decks is loaded into the computer. At some computer installations each user must run his own program. Most large computer systems, however, are run on an "over-the-counter" basis, where each user delivers his deck of cards to a computer operator who loads it into the computer. From the point of view of computer center personnel, the over-the-counter operation is usually preferred, since it leads to a more efficient use of computer time and it prevents inexperienced users from accidentally damaging the computer. However, from the user's point of view there are advantages to a "hands-on" type of use. If the user runs the computer himself, the much shorter time delay between feeding in the program and getting back the results (*turn around time*) permits him to find errors in the program (*debug*) much more rapidly. In addition, when he runs the computer himself, the user can react to unanticipated developments that a computer operator unfamiliar with the program might not know how to deal with. The advantages of a hands-on type of operation are particularly great for the person who is just learning how to write programs. One very effective way to learn programming is the experimental approach of changing a few instructions and seeing what happens when the program is run.

Control Cards

When a program is loaded into the computer in the form of a deck of cards, there are a number of *control cards* which must accompany the program itself. The control cards take the place of the system commands that are typed in by the user in the conversational mode. The number and type of control cards required to run a program are computer-dependent, and therefore this subject will not be discussed here. Just as an example, we show in Figure 1.19 the complete deck

Figure 1.19   Deck of cards needed to run electric field program on an IBM 1130

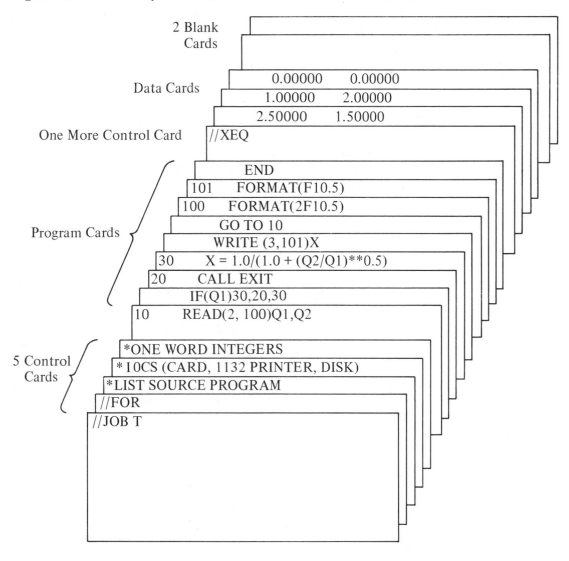

of cards needed to run the program which appeared on page 31 on an IBM 1130 computer. Five control cards precede the program and one follows it. After this last control card there are three data cards. (In this particular program the number of data cards was not specified, since the program has been written to repeatedly read values of Q1 and Q2 from a card, perform a calculation, and write the result, until it reads a card on which Q1 has the value zero.) In addition to the data cards, some computers, such as the IBM 1130, require two additional "dummy" blank cards at the very end of the deck. These two extra cards are not actually read into the computer and are only needed because some card readers do not process the last two cards in the deck.

Punching a Deck of Cards

Several rules must be observed when punching a deck of cards:

1. For numbered FORTRAN statements, the statement number must be punched in columns 1-5 of the card.

2. The FORTRAN statement itself must be punched between columns 7 and 72. If a statement does not fit on a single card, it may be continued on additional cards which have numbers in column 6 to indicate that they are *continuation cards*.

3. All statements must be punched exactly as they appear in the program, including every decimal point and comma. One exception is that extra blank spaces can (almost) always be added to a statement to improve its appearance if this is desired. For example, it is permitted to write

    $X = 1.0$

    in place of

    $X=1.0$

    Since blanks are ignored in FORTRAN, except in certain statements using "literal" constants.

4. The numbers punched on data cards must have a format that is consistent with that specified in the respective FORMAT statement. As an illustration, consider the data cards read by the program on page 31 (see Figure 1.20). According to the READ statement in the program, each data card should contain numerical values for the variables Q1 and Q2. The format of the data should be consistent with the statement:

    100 FORMAT (2F10.5)

    which specifies:

    a. There are two numbers with decimal points on a card.

    b. Each number is contained within a ten column "field" on the card. (The first number must be in columns 1-10, and the second in columns 11-20.)

    c. There are five digits after each decimal point.

Figure 1.20   Format of sample data cards

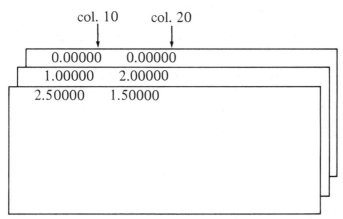

The last rule concerning the placement of the decimal point is actually necessary only if the decimal point is not explicitly punched on the card. If the decimal point is punched on the card, it may appear anywhere within the field without regard to what is specified in the FORMAT statement. It is always a good idea to include an explicit decimal point in numbers that are read in "F" format, as this eliminates the possibility of the decimal point being located improperly. The values for Q1 and Q2 which appear on the last data card of Figure 1.19 are zero, since the program has been written to check for the end of data as signaled by the condition Q1 = 0.0. The last data card could have been left completely blank, and the values read in for Q1 and Q2 would have been read as zero.

Compilation of FORTRAN Programs and Error Messages

Before the computer can execute a program, it first must undergo *compilation*, in which the computer converts the FORTRAN statements of the program into a "machine language" version, in which each FORTRAN statement has been broken up into many elementary operations by means of a master program called a *compiler*. For example, the compiler would reduce the FORTRAN statement

X = 2.0*A**2/B

into a series of elementary operations: multiply A by itself, multiply the result by 2.0, divide *that* result by B, and store *that* result in X. When the computer compiles a FORTRAN program it may uncover errors in the form of some of the FORTRAN statements and print out a message to this effect. If the computer compiles a program without finding any errors, the machine language version of the program is loaded into memory (automatically on many computers). Then the computer attempts to execute the program.

It sometimes happens that due to some error in the program that was not detected during the compilation process, the computer is unable to execute a program. In this case, the computer may report an error message and abort the execution, or it may simply "hang up" and not be able to proceed. Of course, even if the computer does execute the program and produce printed results as expected, there is absolutely no guarantee that the results are correct. There are many types of errors that you may make in writing a program or punching the deck of cards that will not result in any errors that the computer can detect. In a subsequent section on debugging we shall discuss some of the more common types of programming errors, as well as the techniques for locating errors.

In order to become proficient at writing programs and making them work, practical experience is very important for the beginner—even before understanding the finer points. Therefore, before you go on to the next section, punch a deck of cards (such as the one shown in Figure 1.19), and try to run the program on a computer. (If a machine other than an IBM 1130 is used, a different set of control cards will probably be required.) In addition to running the program as it appears on page 31, you should also try changing the program and run the modified versions also. A number of changes are suggested below. Some of the suggestions involve writing completely new programs.

Suggested Modifications to Sample Program

After making each of the suggested modifications, be sure to run the modified program using several sets of data and *check that the results printed out are correct*. If the results appear to be incorrect, try to find and correct any errors in the program. It may happen that the program gives a correct result for some values of the parameters Q1 and Q2 and not others, due to limitations inherent in the algorithm.

1. Run the program using values for Q1 and Q2 other than those specified on the sample data cards of Figure 1.20. Try using values for Q1 and Q2 that are (a) both positive, (b) both negative, (c) one positive, one negative.

2. Modify the program so that statement 30 is based on the formula which gives x in the case of two charges of unlike signs:

$$x = \frac{1}{\left(1 - \left(-\dfrac{q_2}{q_1}\right)^{1/2}\right)}$$

3. Modify the program so that it can be used regardless of the relative sign of the two charges, using the algorithm shown in Figure 1.13.

4. Write a program to find x using the decision tree algorithm shown in a compact form in Figure 1.16. When you run the program try using the same values for Q1 and Q2 that you used in suggestions 1-3. For each pair of values for Q1 and Q2 use several values for ϵ, the parameter that determines the accuracy of the result.

5. Modify the program based on the decision tree algorithm so that it works properly not only when both charges are positive, but also when both are negative, using the method described on page 23.

6. Modify the program based on the decision tree algorithm so that it can be used to find a root of *any* function E(x) in the interval between x = 0 and x = 1, as described on page 23. Run the program using various functions for E(x). For example, you might try

    $$E(x) = \cos x^2 - \sin^2 x.$$

Note that the roots of this function cannot be found analytically.

7. Modify the program based on the decision tree algorithm so that it can be used to find the maximum or minimum of any function in the interval between $x = 0$ and $x = 1$, as described on page 23. Run the program using various functions for $E(x)$. For example, you might try

$$E(x) = xe^{-3x} + x/10.$$

Note that the maximum in this function cannot be found analytically.

8. Write a program using the hill-climbing algorithm, described on page 23, which can be used to find the point in the square region defined by $0 < x < 1, 0 < y < 1$, at which some function $F(x, y)$ has a maximum or a minimum. For example, you might try

$$F(x, y) = (3x - 1)^2(5y - 1)^2,$$

for which the maximum *can* be found analytically.

9. Modify the hill-climbing algorithm, so that it can be used to find the maximum or minimum of any function $F(x, y)$ *anywhere* in the xy-plane, not just in the unit square (or some other specified region). In the original hill-climbing algorithm, the step sizes Δx and Δy are halved each step (see Figure 1.18). In the modified hill-climbing algorithm, adopt the following procedure: double the step size Δx if this step is in the same direction as the last (e.g., if $x \leftarrow x + \Delta x$ is used in both steps), and halve the step size Δx if this step is in the opposite direction as the last (i.e., if $x \leftarrow x + \Delta x$ is used in one step and $x \leftarrow x - \Delta x$ in the other). The same procedure can be followed for the step size in the y-direction, Δy. The effect of this procedure is continually larger steps in the same direction (for either x or y) until we overshoot the mark, and then continually smaller steps. Write a program based on this algorithm, and run it using several functions for $F(x, y)$.

1.6 Additional Features of FORTRAN IV

FORTRAN IV has a number of important features in addition to those already discussed. While many simple programs can be written without using these features, they can prove extremely useful in certain cases.

COMMENT Statements

A COMMENT statement, unlike all other types of statements, is not an instruction to the computer. Its only purpose in a program is to explain something to anyone who reads the program. Even if no one else will be looking at your program, it is very convenient to use many COMMENT statements so that when you look at the program later you can more easily remember the program's purpose and content. A COMMENT statement is designated by the letter "C" in column 1 of a card. The remainder of the card may contain anything, as indicated in the following example:

```
col. 1
  ↓
  C              ADD A AND B, AND STORE THE RESULT IN X.
  C
        X = A + B
```

Real and Integer Variables and Constants

In FORTRAN a distinction is made between numbers which contain a decimal point—*reals* or *floating point numbers*, and numbers which do not—*integers* or *fixed point numbers*. Examples of integers and reals are shown below:

integer	real
−546	3.14159
9999	−1.0
0	2.31 E-9

The last real in this example illustrates the use of power of ten notation in FORTRAN: 2.31 E-9 represents the number 2.31×10^{-9}.

The distinction between integers and reals is important because arithmetic operations are carried out differently in the two cases. In the case of integers, each arithmetic operation yields a result which is truncated to the nearest lower integer. This is known as *integer arithmetic* or *fixed point arithmetic*. In the case of reals, arithmetic operations are performed in the usual way according to *real arithmetic* or *floating point arithmetic*. For example, the result of the division 7/2 would be 3 according to the rules of integer arithmetic, whereas the result of 7.0/2.0 would be 3.5 according to the rules of real arithmetic. The loss of precision that occurs when integer arithmetic is used is not always undesirable and sometimes can be used very advantageously.

Variables, like numerical constants, can be divided into real and integer categories. FORTRAN has two ways to place a variable in the desired category. At the beginning of a program we can specify whether we wish a variable to be considered real or integer using a *declaration statement* such as

```
REAL X, Y, Z
INTEGER A, B, C
```

The variables X, Y, and Z are treated as reals and the variables A, B, and C as integers. With the second way it is not necessary to explicitly include every variable in a program in a declaration statement. We simply follow this FORTRAN convention: variables with names beginning with one of the letters I, J, K, L, M, N are integer variables, and variables with names beginning with one of the other letters of the alphabet are real variables. Apart from this restriction, variable names may consist of an arbitrary sequence of letters and numbers, provided that the name begins with a letter and does not have more than some maximum length (five characters on some computers, six on others). To illustrate, we list below some legitimate variable names. The classification into real and integer types has been made on the basis of the FORTRAN convention, assuming that none of the variables appears in a declaration statement which alters its type.

integer variables	real variables
N, K24Q, KOUNT, LIM8, JJJJ	X, Q1, SUM, A23, FUNCT

All the variables and constants which appear in an assignment statement to the right of the equal sign must be of the same type, either all real or all integer. *
The four possibilities consistent with this rule are shown in the following table:

variable to left of equal sign	expression to right of equal sign
1. real	real
2. integer	integer
3. real	integer
4. integer	real

Each of these cases is illustrated by an example:

1. *real = real:*

$$C = 5.0*A/B$$

If this statement were to appear in a program after A and B have been assigned values of 2.0 and 3.0, respectively, then the execution of this statement causes C to be assigned a value 3.3333333. (The exact number of significant digits depends on the word size of the particular computer.)

2. *integer = integer:*

$$K = 2*(L/2)/3$$

If L has been previously assigned a value of 5, then the execution of this statement causes K to be assigned a value of 1 (obtained by the following sequence of integer arithmetic operations: $5/2 = 2$, $2 \times 2 = 4$, $4/3 = 1$).

3. *real = integer:*

$$X = 5*K/2 + 1$$

If K has been previously assigned a value of 1, then the execution of this statement causes X to be assigned a value of 3.0. Note that the expression is first evaluated according to integer arithmetic and then it is converted to a floating point (real) number.

4. *integer = real:*

$$K = 0.5*A**2$$

---

* Some computers do permit violations of this rule, but it is best to avoid the use of "mixed" expressions.

If A has been previously assigned a value of 3. 0, then the execution of this statement causes K to be assigned a value of 4. Note that the expression is first evaluated according to floating point arithmetic and then the result is truncated to the next lower integer. This example illustrates the one exception to the previously mentioned rule that all variables and constants appearing to the right of an equal sign must be of the same type: integral exponents can (and should) be written without a decimal point.

Arrays and the DIMENSION Statement

It is sometimes convenient when dealing with a number of related variables not to give each one a different name, but to distinguish them by an index, which is analogous to the use of subscripts in mathematical notation. In FORTRAN indices must be enclosed in parentheses. For example, the following statements might appear in a program:

Q(1) = 2. 0
Q(2) = 5. 2

The two variables Q(1) and Q(2) are called the *elements* of the *array* Q. (Note that two variables named Q1 and Q2 do *not* constitute elements of any array and they have no relation to either the array Q or each other.)

Each element of an array corresponds to a different location in the computer memory. Normally the elements of an array are stored in a contiguous *block* of memory locations. Thus, the index in parentheses refers to a particular memory location within the block. For this reason, the index must always be a positive integer that does not exceed the total number of memory locations within the block reserved for the particular variable. The number of locations to be reserved for each array must be specified in a DIMENSION statement, usually placed near the beginning of a program. As an example, consider the DIMENSION statement:

DIMENSION X(10), Q(5), Z(1000)

This statement specifies that 10 memory locations should be reserved for the array X, 5 for the array Q, and 1000 for the array Z. In this case, it would therefore be perfectly legitimate to have the following array elements referred to in a program which contained this DIMENSION statement: X(8), Q(5), Z(982). However, any references to X(11), Q(9), Z(1027), X(0), Q(-2), or Z(3. 64) would not be legitimate. The indices used to refer to an element of an array may be variables or mathematical expressions as well as constants. For example, the following are completely valid statements:

X(J) = 2
X(2*K + 1) = 10

provided that J and K have been assigned numerical values in some previous statements and that the values of the expressions in parentheses is within the range specified by the appropriate DIMENSION statement. On many computers, the most complicated expression permitted for the index of an array is $c_1*v +$ c_2, where c_1 and c_2 represent two fixed point constants and v represents a fixed point variable. Thus, for example, the following statement would not be permitted:

$$X(K**2 + 2*K) = 10$$

Just as variables in mathematics may have more than one subscript, variables in FORTRAN may have more than one index. The number of indices specifies the number of dimensions of the array. For example, the elements of a two-dimensional array Y might include Y(1, 1), Y(1, 6), Y(2, 3), and Y(3, 5). The size of arrays of more than one dimension must also be specified in a DIMENSION statement. The statement

$$\text{DIMENSION } Y(4, 10)$$

specifies that the first index of the two-dimensional array Y can take on values 1, 2, 3, 4, and the second index can take on values 1, 2, ..., 10. The statement therefore specifies that a block of $4 \times 10 = 40$ memory locations be reserved for the array Y. One-dimensional arrays are sometimes called "vectors" and two-dimensional arrays are called "matrices." On many computers arrays of more than three dimensions are not permitted.

DO Statement

We have seen earlier how a loop in a program can be created using the GO TO and IF statements. An alternative method is to use to the DO statement. We can illustrate the meaning of the DO statement by an example:

$$\text{DO } 5 \text{ K} = 7, 87$$

This statement instructs the computer to execute repeatedly all the statements that follow, up to and including statement number 5, a specific number of times, setting the *loop index* K equal to 7 initially and increasing it by one each time through the loop, until it exceeds 87 which is the last time through the loop. (Thus the loop is to be executed 81 times.)

From the meaning of the DO statement, it should be clear that the following two sets of instructions are equivalent:

```
C      EXAMPLE OF A LOOP           C      EXAMPLE OF A LOOP

C      USING AN IF STATEMENT       C      USING A DO STATEMENT

C                                  C
       N = 6                              DO 2 N = 7, 100
   1   N = N + 1                           .
         .                                 .
         .                                 .
         .
                                       2   CONTINUE
       IF(N − 100)1, 1, 2                   .
   2   CONTINUE                            .
         .                                 .
         .
```

To illustrate how a DO statement might be used in a program, we show a program which reads in numerical grades for some number of students and computes an average grade. As indicated in this example, the upper (or lower) limit on the loop index specified in the DO statement may be a variable as well as a constant. Many COMMENT statements have been used to explain the role of the statements in the program.

```
C          READ A NUMERICAL VALUE FOR N, THE NUMBER OF GRADES.
           READ(2, 100) N
C          SET THE SUM OF THE GRADES TO ZERO INITIALLY.
           SUM = 0.
C          REPEAT ALL INSTRUCTIONS UP TO STATEMENT NUMBER 10
               FOR J = 1, 2, . . . , N.
           DO 10 J=1, N
C          READ A DATA CARD CONTAINING A GRADE.
           READ(2, 101) GRADE
C          ADD THIS GRADE TO THE SUM.
    10     SUM = SUM + GRADE
C          LET FN BE THE FLOATING POINT EQUIVALENT OF N.
           FN = N
C          COMPUTE THE AVERAGE GRADE.
           AVG = SUM/FN
C          PRINT THE RESULT.
           WRITE(3, 102) AVG
C          STOP THE PROGRAM.
           CALL EXIT
   100     FORMAT(I10)
   101     FORMAT(F10. 5)
   102     FORMAT(19H THE AVERAGE GRADE=, F10. 5)
           END
```

If a loop defined by a DO statement contains one or more IF statements, the program may execute the loop fewer times than is specified in the DO statement. An example of such a premature exit from the loop is given in the program below. Note that when a premature exit occurs, the latest value of the loop counter is saved and may be used in another part of the program.

```
C          THIS PROGRAM WAS WRITTEN TO READ 100 CARDS, EACH
C          CONTAINING A NUMBER, AND TO INDICATE WHICH IS
C          THE FIRST CARD OUT OF SEQUENCE (THAT IS, THE
C          FIRST CARD THAT HAS A NUMBER SMALLER THAN THE
C          PRECEDING CARD).
C
           DIMENSION NUM(100)
           READ(2, 100) NUM(1)
           DO 10 J = 2, 100
           READ(2, 100) NUM(J)
C
C          THE IF STATEMENT CAUSES A JUMP OUT OF THE
C          LOOP (TO STATEMENT 20) IF NUM(J) IS LESS
C          THAN NUM(J-1), THE PREVIOUS NUMBER.
```

```
        IF(NUM(J) — NUM(J-1)) 20, 10, 10
   10   CONTINUE
        GO TO 30
   20   WRITE(3, 100) J
   30   CALL EXIT
  100   FORMAT(I10)
        END
```

In many programs a loop may be contained within another loop. Such *nested loops* can easily be created using DO statements as indicated in the following example:

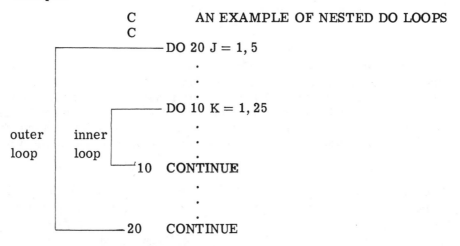

C　　　　　　　AN EXAMPLE OF NESTED DO LOOPS
C

DO 20 J = 1, 5

DO 10 K = 1, 25

outer loop　inner loop

10　CONTINUE

20　CONTINUE

In this example, the computer is instructed to execute the outer loop 5 times for values of the loop index $J = 1, 2, \ldots, 5$. Each time through the outer loop, it is instructed to execute the inner loop 25 times for values of the loop index $K = 1, 2, \ldots, 25$. Thus, the inner loop is executed a total of $5 \times 25 = 125$ times. Any number of loops may be contained within a loop; this includes the possibility of loops within loops within loops.

Some possibilities are shown schematically in Figure 1.21; (c) indicates that nested loops may have a single last statement in common.

Figure 1.21　Possibilities for nested DO loops

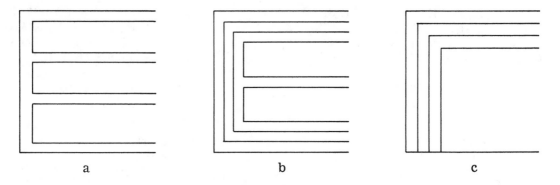

a　　　　　　　　　b　　　　　　　　　c

An ambiguous structure such as the one shown below is not allowed, since neither loop is contained within the other.

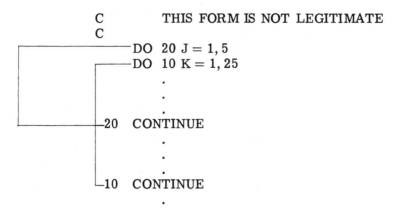

```
C              THIS FORM IS NOT LEGITIMATE
C
       DO  20 J = 1, 5
       DO  10 K = 1, 25
            .
            .
            .
20     CONTINUE
            .
            .
            .
10     CONTINUE
            .
            .
            .
```

Using the loop index to the left of an equal sign anywhere within a loop is also
not allowed, since this redefines its value and the loop is not executed the
proper number of times. For a similar reason the variable names used as loop
indices in nested DO loops must be different. However, instructions outside of
a loop may use the variable name used as the loop index in some different way.
In the following examples, (a) and (b) are illegitimate DO loops, while (c) and (d)
are allowed.

```
C        AN IMPROPER USE OF THE
C        LOOP INDEX WITHIN A DO LOOP
     DO  10 J = 1, 3
          .
          .
          .
     J = 1
10   CONTINUE



              (a)
```

```
C        AN IMPROPER USE OF THE
C        SAME LOOP INDEX IN NESTED
C        DO LOOPS
     DO  10 J=1, 5
          .
          .
          .
     DO 6 J = 1, 3
          .
          .
          .
6    CONTINUE
10   CONTINUE
              (b)
```

```
C        IT IS OK TO USE THE SAME
C        VARIABLE AS A LOOP INDEX
C        AS LONG AS THE LOOPS ARE
C        NOT NESTED.
C
     DO  10 J = 1, 10
          .
          .
10   CONTINUE
     DO  20 J = 1, 30
          .
          .
20   CONTINUE
              (c)
```

```
C        IT IS OK TO USE THE LOOP
C        COUNTER VARIABLE NAME IN
C        SOME OTHER WAY IN INSTRUC-
C        TIONS OUTSIDE THE LOOP.
C
     DO  10 J = 1, 10
          .
          .
10   CONTINUE
     J = 1
          .
          .
          .
              (d)
```

Mathematical Functions

One method of taking the square root of a quantity in FORTRAN is to raise the quantity to the $\frac{1}{2}$ power, as in the statement

Y = X**0. 5

The FORTRAN language provides an alternative method of instructing the computer to calculate a square root using the square root function designated by the name SQRT. Thus, the statement

Y = SQRT(X)

accomplishes exactly the same result as the previous one.

The square root is one of a number of functions that are "built-in" to the FORTRAN language. The following list includes some of the more commonly used functions and their FORTRAN names, but it is by no means complete.

Function	FORTRAN Name
$\sqrt{x}$	SQRT(X)
sin x	SIN(X)
cos x	COS(X)
tan x	TAN(X)
arctan x	ATAN(X)
$\log_e x$	ALOG(X)
$|x|$	ABS(X)
e^x	EXP(X)

For the three trigonometric functions SIN, COS, and TAN, the argument X must be expressed in radians, not degrees. The value of ATAN(X) is computed in radians.

As indicated earlier, a computer is capable of performing only very simple arithmetic operations, which do not include these mathematical functions. When the computer encounters one of these special function names, it temporarily stops the execution of the program and transfers to a "subprogram" or "subroutine," which consists of a set of instructions that enables the computer to calculate the value of the function for the specified value of the argument, usually using a series expansion involving some finite number of terms. (Other recursive methods may sometimes be used in place of a series expansion to evaluate certain functions.) Usually, enough terms in the expansion are used so that the accuracy of the result is limited only by the finite word size of the computer. The standard mathematical functions are called built-in because each of the subprograms can be called out of a program library by simply using the appropriate code (SQRT, SIN, etc.) in a FORTRAN program. It is also possible in FORTRAN for the user to write his own special purpose subprograms that can be called out of a program library in a similar manner, although we will not discuss this here.

FORMAT Statements

A FORMAT statement, as noted previously, may be referred to by one or more

READ or WRITE statements. The FORMAT statement contains a string of *format specifications* separated by commas, for example,

100   FORMAT(F11.5, E15.5, I6, 29H ARE THE VALUES OF A, B, and N)

This statement contains four types of format specifications:

F format, for floating point variables,
E format, for floating point variables, expressed in exponential form,
I format, for integer (fixed point) variables,
H format, for specific strings of characters of any length, used to print labels or titles on the output.

The meaning of the four format specifications in the sample **FORMAT** statement is as follows:

F11.5   specifies that the quantity to be read or written is a floating point number occupying 11 spaces and has five digits after the decimal point, for example,   —45.12347 (In this case, two blank spaces must be added in front of the number, in order that it occupy a total of 11 spaces counting the sign and the decimal point.) If the same number were written in F6.2 format it would appear as —45.12, with no leading blanks. Finally, if we attempted to write the number in F5.2 format, we might get   ***** (It can't be done.)

E15.5   specifies that the quantity to be read or written is a floating point number in exponential (power of ten) form. It occupies a total of 15 spaces with 5 digits after the decimal point, for example,   0.8999E-12 (0.8999×10^{-12}). Note that in this case four blank spaces must appear in front of the number in order that it occupy a total of 15 spaces.

I6   specifies that the quantity to be read or written is an integer which takes up 6 spaces, for example, —12345

29H ARE THE VALUES OF A, B, AND N   The "29H" specifies that the 29 characters: "ARE THE VALUES OF A, B, AND N" are to be written exactly as they appear. Unlike the three preceding formats, the H format does not refer to any variable that appears in a READ or a WRITE statement; it is simply used to write out a specific string of characters. On some computers the characters to be written need only be enclosed in single quotation marks, avoiding the need for a character count, i.e., 29 in this case.

The following program illustrates the use of the F, E, I, and H format specifications.

```
C          ILLUSTRATION OF A FORMAT STATEMENT
C          USING F, E, I, and H FORMAT SPECIFICATIONS
C
           A = 4.0/3.0
           B = 25000.
           N = 67
           WRITE(3, 100) A, B, N
100        FORMAT(F11.5, E15.5, I6, 29H ARE THE VALUES OF A, B, AND N)
           CALL EXIT
           END
```

When this program is executed, the values of A, B, and N will be written out on device number 3, according to the format specifications in FORMAT statement 100. Assuming that device number 3 is the printer, the following line would be printed:

1.33333   0.25000E+5   67 ARE THE VALUES OF A, B, AND N

As another example, suppose the FORMAT statement in the program were replaced by

100   FORMAT(30H1   A   B   N//F11.5, E15.5, I6)

The first format specification in this FORMAT statement indicates that the 30 characters (including blanks) which follow the "30H1" are to be printed exactly as they appear (this includes everything up to the two slashes). The "1" functions as a *control character*. There are three useful control characters, which are used to control the way a line is printed out on a printer, typewriter, or teletypewriter:

control character	meaning
(blank)	Print this line and advance to the next.
1	Skip to a new page, then print this line and advance to the next.
+	Print this line but don't advance to the next.

Thus, in the FORMAT statement we are presently considering, the computer is instructed to advance the printer to a new page and then print the heading: "   A   B   N" The two slashes (//) which come after the 30H format instruct the computer to skip two lines before printing anything further. Thus, the actual numerical values for A, B, and N appear on the third line after the heading:

A   B   N

1.33333   0.25000E+5   67

Note that if the numerical value of A had actually needed the alloted eleven positions, the leading digit "1" would be the first character on the line and would cause the printer to (unintentionally) skip to a new page.

There are other types of formats in addition to the F, E, I, and H; we shall discuss one of them in the next section.

DATA Statements and the A Format Specification

We have previously seen two methods by which variables in a FORTRAN program can be given numerical values: the READ statement and the assignment statement. The DATA statement provides a third alternative. An example of a DATA statement is

DATA A, B, C, KOUNT/1.0, 2.0, 5.7, 13/

which is equivalent to the four assignment statements:

$$A = 1.0$$
$$B = 2.0$$
$$C = 5.7$$
$$KOUNT = 13$$

There is, however, one difference: the DATA statement(s) must always be placed at the beginning of a program, whereas the assignment statements may appear anywhere.

DATA statements can be used to define nonnumerical quantities as well as numerical ones. If we have the following DATA statement in a program:*

DATA X, Y, Z, DOG/ 'B', '+', ' ', 'FLEA'/

then during execution, the letter "B" is stored in the variable X, the character "+" is stored in the variable Y, the character " " (blank) is stored in the variable Z, and the sequence of letters "FLEA" is stored in the variable DOG. (The maximum number of characters that can be stored in one variable usually ranges from four to six depending on the word size of the computer.) If letters or other characters are stored in a variable it does not make any sense to use the variable in performing arithmetic operations (e.g., it makes no sense to add the letter "B" to the character "+"). This type of variable can, however, be used in assignment statements which do not involve any arithmetic operations, as in the following example:

DATA DOG/'FLEA'/
CAT = DOG

The DATA statement instructs the computer to store the letters "FLEA" in the variable DOG. The assignment statement then stores the value of DOG (which is "FLEA") in the variable CAT.

In order to read or write quantities consisting of strings of characters we may use the A format, as shown in the following example:

```
        DATA DOG/'FLEA'/
        CAT = DOG
        WRITE(3, 110)CAT
110     FORMAT(A4)
        CALL EXIT
        END
```

During execution of this program the WRITE statement specifies that the value of the variable CAT ("FLEA") be printed in A4 format, as indicated in FORMAT statement 110. But since the first character on a line is assumed to be a control character, the F in "FLEA" is interpreted as a control character, and the printed message consists of the remaining characters "LEA". It is desirable, there-fore, to have the printer skip one or more spaces on the line before the first

---

*   On some computers, it is necessary to use the H format specification instead of the single quotation marks, in which case this statement would be written

DATA X, Y, Z, DOG/ 1HB, 1H+, 1H, 4HFLEA/

character appears. For example, if we want the printer to skip the first 10 spaces on a line and then print the characters "FLEA" we need only modify the FORMAT statement to read:

110   FORMAT(10X, A4)

Thus, the X format is used to skip some number of blank spaces. The A format is used to print characters that are stored in a specific variable, in contrast to the H format used to print characters on titles not associated with any specific variables in the computer memory.

Precision of Computations

Due to its finite word size, the computer can not perform floating point arithmetic operations and obtain exact results. For example, if the computer has a word size equivalent to eight significant figures, then the value computed for the quotient 7.0/2.0 is only correct to eight figures. Note that even though the exact answer   3.50000... has all zeros after the second figure, an inaccuracy arises, since on most computers numbers are represented internally using the binary rather than the decimal system. Thus most numbers (base 10) suffer a loss of precision (round-off error) when the computer represents them in binary (base 2) form, even before the numbers are used in any arithmetic operations. In most computations, the loss of precision means that the final result may be incorrect after the seventh or eighth figure, which may be considered only a minor annoyance. However, in some cases far more serious errors can occur. Random accumulation of round-off error due to a large number of arithmetic operations can cause the final result to be inaccurate by a significant amount. Even a single arithmetic operation can cause a large loss of precision, for example, if two numbers of nearly equal size are subtracted from one another.

On most computers it is possible to specify that the number of significant digits for each arithmetic operation be increased from its normal value. To accomplish this, the computer stores each quantity in two or possibly three words of memory, permitting it to double or triple the number of significant digits (*double precision* or *triple precision*). If you suspect that the results of a program are inaccurate due to round-off error, then it is advisable to run the program again using double or triple precision (usually accomplished by adding the appropriate control card). If you find that the numerical results differ substantially between the original (single precision) run and the double precision run, then in fact round-off error is present. While the double precision result is certainly more trustworthy than the single precision one, it too may be in error if the two results are substantially different.

In discussing the precision of numbers, we have so far been considering their internal representation in the computer. The number of figures in a printed result is, of course, determined by the appropriate FORMAT statement. On a computer having a word size equivalent to eight figures, the (single precision) result of the quotient 7.0/2.0 printed out to four, eight, and twelve figures might be, respectively,

3.500   3.4999999   3.49999992174

In this example the figures after the eighth are meaningless, and when fewer than eight figures are printed, the least significant figure is rounded off according to the value of the next figure.

Debugging FORTRAN Programs

Most programmers find that their programs do not work properly the first time they are run. In fact, it is not at all unusual for the debugging of a program to require considerably more time and effort than was required to write the program in the first place. Sometimes the final correct version of the program bears little resemblance to the original version! This is largely due to the many different kinds of errors that can be made in writing a program. While the computer can detect some types of errors in your program and alert you to them, there are other kinds of errors (the most difficult to find), which do not give any obvious indication of anything wrong—except that the answer is not correct. For this reason it is important to take a very skeptical attitude towards the results of a computer calculation, checking it whenever possible by a hand calculation. In cases where this is impractical, check for the following:

Internal consistency—often calculations can be done more than one way, do the answers obtained using different methods agree?

Reasonableness—often by making simplifying assumptions, an estimate to the answer can be found; how does the result compare with the estimate?

Limiting cases—when a calculation is done involving parameters that must be assigned numerical values, sometimes the result is obvious for certain values of the parameters; are the results for these limiting cases correct?

It cannot be overemphasized that you should not grant automatic validity to the results of a computer calculation—check them in some way. If you suspect that an error is present in a program, the general approach in trying to locate the error is to go through the program step by step to see how the computer obtained its results. The idea is to assume nothing, but simply follow each instruction mechanically. In some cases the error may be due to something trivial, like a mispunched card (perhaps you accidentally typed the letter "O" when you intended to type the number "0"), or possibly you inadvertently left out a card. Some of the more subtle kinds of errors involve the misuse of certain FORTRAN statements, some of which have already been discussed.   The more common errors fall into one of the following categories:

1. Use of an array index outside the range specified in a DIMENSION statement

This very common mistake can occur in a number of ways, one of which is illustrated below:

        J = X**2*SIN(Y/Z)
        F(J) = 1.0
            .
            .
            .

Whenever an integer variable evaluated from some mathematical expression is used as an array index, we must be very careful that the form of the expression never causes the index to go outside its permitted range. In the present example, we are in trouble if Y and Z assume values that make SIN(Y/Z) negative or zero.

2. Infinite loops

An infinite loop will occur any time the exit from a loop depends on some condition which is never satisfied. An example of an infinite loop is shown in the following program which is supposed to calculate and print a table of values of the function e^x for $x = 0, 1, 2, \ldots, 10$. When the variable X is incremented by 1.0 each time through the loop, the result is only accurate to some specific number of significant figures. Thus, on the tenth time through the loop the value of X might be 9.9999999351... instead of exactly 10.0, so that the IF statement never causes the program to branch out of the loop to statement 2.

```
C          AN EXAMPLE OF AN INFINITE LOOP
C
           X = 0.
   1       E = EXP(X)
           WRITE(3, 101)X, E
           X = X + 1.0
           IF(X − 10.0)1, 2, 1
   2       CALL EXIT
 100       FORMAT(2F10.5)
           END
```

One solution is to change the IF statement in this program to

$$\text{IF(ABS(X − 10.0) − 0.0001) 2, 1, 1}$$

which would cause a branch out of the loop when X differs from 10.0 by no more than .0001.

3. Failure to bypass the unselected branch

As an example of this error, we return to our sample problem. To compute the coordinate of the point at which the net electric field from two point charges (Q1 and Q2) vanishes, we use one of two formulas, depending on whether the ratio Q1/Q2 is positive or negative.

```
C          AN EXAMPLE OF A FAILURE TO BYPASS THE
C          UNSELECTED BRANCH
C
           .
           .
           .
           IF(Q2/Q1)2, 1, 1
   1       X = 1.0/(1.0 + (Q2/Q1)**0.5)
   2       X = 1.0/(1.0 − (−Q2/Q1)**0.5)
   3       CONTINUE
           .
           .
           .
```

If the value of Q2/Q1 happens to be negative, the correct result is obtained. If the ratio is positive, the program branches to statement 1, where X is calculated properly, but then it recomputes X according to the wrong formula (statement 2). In order to bypass the unselected branch we must insert the statement

```
       GO TO 3
```

between statements 1 and 2.

4. *Division by zero*

If this impossible operation is attempted in a FORTRAN program, the result depends on the particular computer: some computers cause an error message to be printed out, but many do not. If no error message is printed, the computer may take the value to be some arbitrary number (possibly zero, one, or the largest number that can be defined on the particular computer) in order not to terminate the program. One way to avoid a division by zero is to test the divisor (using an IF statement) before performing the operation. For example, if in the previous program, values for Q1 and Q2 are read in on a data card, before computing the ratio Q2/Q1, check that Q1 is not zero by inserting the following statements before the IF statement in the program:

```
        IF(Q1) 20, 10, 20
10      CALL EXIT
20      CONTINUE
```

In this case, if the value read in for Q1 is zero the program would terminate. (This serves a dual purpose in the present case. In addition to avoiding a division by zero, we can use this check to terminate the program after the desired data cards have all been read by including an extra card on which Q1 has the value zero.)

5. *Argument of a function outside the allowed range*

The most obvious example is attempting to use the square root function SQRT with a negative argument. Other functions, such as the exponential, EXP, also have limitations on the range of values the argument may assume. (In the case of the exponential function, the range is dictated by the largest (or smallest) number that can be represented on the particular computer.) As in the case of a division by zero, many computers do not print an error message if the argument of a function is outside the allowed range.

6. *Data not read properly*

Given the complexities of the rules concerning FORMAT statements, it sometimes happens that the format of numbers on a data card does not exactly agree with that specified in the relevant FORMAT statement. In such a case, the computer may read in incorrect values for certain quantities. The best way to guard against this possibility is to use an "echo check" in a program, which means to print out the input quantities right after they are read in, before any calculations are done using them.

2

Electrostatics and Electric Circuits

In this chapter we investigate a number of problems in the areas of electrostatics and electric circuits. In section 2.1 we make use of a technique for generating graphic displays on the printer; this technique will be further extended in Chapter 3. In the remaining sections we discuss some simple techniques for numerical integration and for the numerical solution of differential equations.

2.1 Equipotential Plots for Two Point Charges

Given a number of point charges $q_1, q_2, \ldots, q_n$, the electrical potential at any point in the xy-plane can be obtained from

$$V(x, y) = \sum_{j=1}^{n} \frac{q_j}{r_j},\qquad(2.1)$$

where r_j is the distance from the charge q_j to the point (x, y) and the Coulomb force constant $k = \dfrac{1}{4\pi\epsilon_0}$ has been omitted. For the case of two charges we have

$$V(x, y) = \frac{q_1}{r_1} + \frac{q_2}{r_2}.\qquad(2.2)$$

If the potential $V(x, y)$ is found at all points in the xy-plane, a family of *equipotential curves* can be obtained by connecting all points having the same values of the potential. In practice, we can calculate $V(x, y)$ at some finite grid of points (see Figure 2.1), and then approximately draw equipotential curves. A computer program for carrying out such a procedure is described in the next section. In a modified version of the program, the computer actually draws the equipotential curves.

Equipotential Plotting Program (version 1)

The first version of this program is given in flow diagram form in Figure 2.2 and listed on page 56. The program first reads from a data card numerical values for the parameters A1, A2, Q1, and Q2, where

A1 = x-coordinate of charge q_1
A2 = x-coordinate of charge q_2
Q1 = value of charge q_1
Q2 = value of charge q_2.

It is assumed that the y-coordinates of both charges are zero.

Fig.2.1   Grid of 21 × 21 points at which potential V(x, y) is found

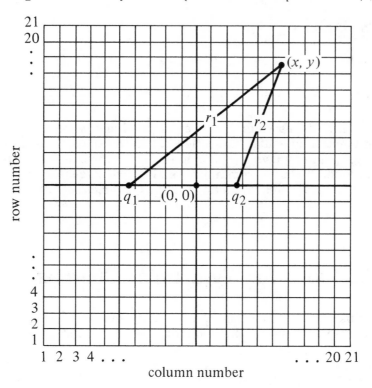

After the data card has been read, the program proceeds to calculate the electrical potential V at all lattice points within the square defined by $-10 \leqslant x \leqslant +10$, $-10 \leqslant y \leqslant +10$. The points are identified by a pair of indices (j, k), where $j = 1$, $2, \ldots, 21$ indicates the row number, and $k = 1, 2, \ldots, 21$ indicates the column number. At each lattice point, starting with $(j, k) = (1, 1)$, the program first determines the coordinates (x, y) in terms of the indices (j, k), using

$$x = k - 11$$
$$y = j - 11.$$

It then computes the distances r_1 and r_2 from the point (x, y) to each of the charges which are located at the points $(A1, 0)$ and $(A2, 0)$ (see Figure 2.1). We add .000001 to r_1 and r_2, so that in computing the potential V according to equation 2.2, the computer never divides by zero. After finding the potential at one point, the program proceeds to the next point within the row. Once a row is complete, the array $V(K), K = 1, 2, \ldots, 21$, which contains the values of the potential at each of the 21 points within the row, is printed out on a single line. The program then treats the next row of points in exactly the same manner, printing the array $V(K), K = 1, 2, \ldots, 21$, for the next row of points on the line below the previous one. After it prints out the potential for all 21 rows, the process is complete, and the program reads the next data card (if any), containing a new set of parameters A1, A2, Q1, Q2.

Figure 2.2   Flow diagram for equipotential plotting program (version 1)

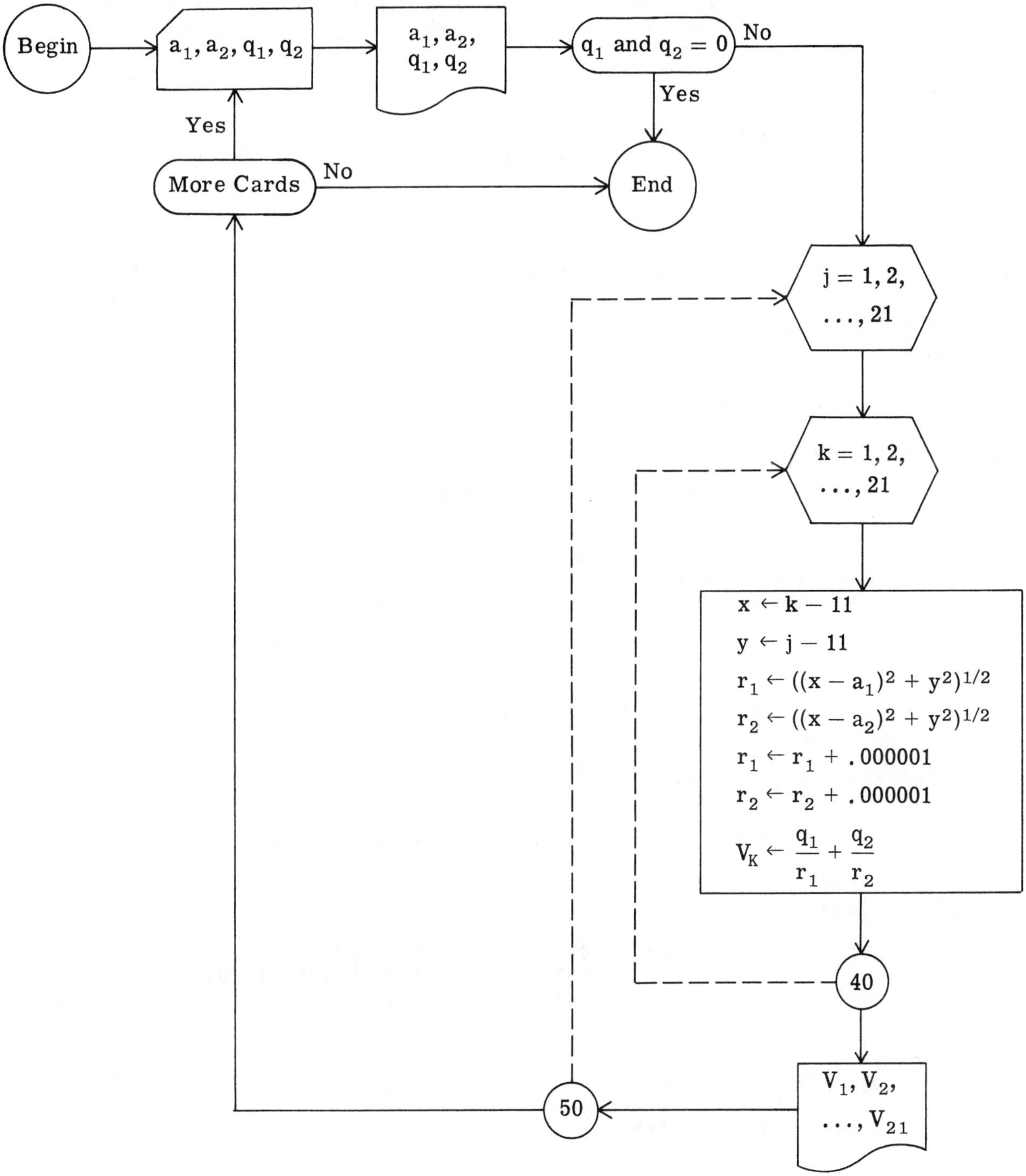

```
C                 ELECTRICAL POTENTIAL FOR TWO POINT CHARGES
C
C.................................................................
C
C                 THIS PROGRAM CALCULATES THE ELECTRICAL POTENTIAL FOR 2 POINT
C                 CHARGES, AT AN ARRAY OF 21 X 21 POINTS IN THE RANGE X=-10 TO X=+10
C                 AND Y=-10 TO Y=+10.
C
      DIMENSION V(21)
C
C                 READ A DATA CARD CONTAINING A1, A2, Q1 AND Q2.
C                 A1 AND A2 ARE THE X COORDINATES OF CHARGES Q1 AND Q2.
C
   10 READ(2,1000)A1,A2,Q1,Q2
      WRITE(3,1001)A1,A2,Q1,Q2
C
C                 IF THE DATA CARD READ IN IS BLANK, THEN Q1 AND Q2 WILL BE ZERO
C
      IF(Q1)30,20,30
   20 IF(Q2)30,60,30
   30 CONTINUE
C
C                 THIS LOOP GOES OVER 21 ROWS.
C
      DO 50 J=1,21
C
C                 THIS LOOP GOES OVER THE 21 COLUMNS  WITHIN A ROW.
C
      DO 40 K=1,21
C
C                 THE X AND Y COORDINATES ARE CALCULATED FROM THE ROW AND COLUMN
C                 INDICES J AND K.
C
      X=K-11
      Y=J-11
C
C                 CALCULATE THE DISTANCES R1 AND R2, FROM THE POINT (X,Y) TO EACH
C                 OF THE TWO POINT CHARGES.
C
      R1=SQRT((X-A1)**2+Y**2)
      R2=SQRT((X-A2)**2+Y**2)
C
C                 THE NEXT TWO STATEMENTS ARE JUST IN CASE R1 OR R2 ARE ZERO.
C
      R1=R1+.000001
      R2=R2+.000001
C
C                 CALCULATE THE NET ELECTRICAL POTENTIAL FOR THE 2 POINT CHARGES.
C                 WE MUST STORE THE POTENTIAL FROM EACH OF THE 21 POINTS IN A
C                 ROW, BEFORE WRITING THESE VALUES OUT.
C
      V(K)=Q1/R1+Q2/R2
   40    CONTINUE
   50 WRITE(3,1002)(V(K),K=1,21)
C
C                 READ NEXT DATA CARD.
C
      GO TO 10
 1000 FORMAT(4F10.5)
 1001 FORMAT(4H1A1=,F10.5,4H A2=,F10.5,4H Q1=,F10.5,4H Q2=,F10.5)
 1002  FORMAT(21F5.2//)
   60 CONTINUE
      CALL EXIT
      END
```

Comments on Data Cards and Sample Results (version 1)

The output shown in Figure 2.3 was obtained using a data card with the following numerical values for the parameters:

A1	A2	Q1	Q2
—3.0	3.0	2.0	—1.0

A number of equipotential curves have been drawn by hand directly on the computer output. To draw one such curve, for example, the one for V = 0.10 (drawn thickest), it is necessary to find, by interpolation, the approximate location of points for which V = 0.10, and then draw a smooth curve which passes through these points.

Figure 2.3

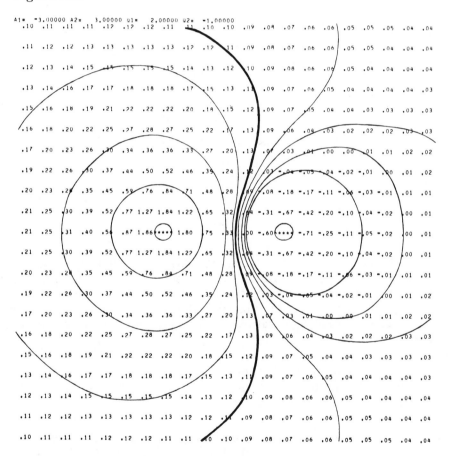

To illustrate the interpolation procedure, suppose the sixteen values of the potential shown in Figure 2.4 appear in some part of the output of Figure 2.3. The sixteen points having these potentials may be assumed to lie at the intersections (*nodal points*) of an xy-coordinate grid drawn through the decimal points. Points for which V = 0.10 can be located approximately on the grid lines by a linear interpolation. These points can be found by eye and are shown as asterisks in Figure 2.4. An equipotential curve can then be obtained by drawing a smooth curve through the asterisks. Following the same procedure for other values of the potential, we can obtain (with considerable effort) a set of equipotential curves.

Figure 2.4   Interpolation procedure for drawing an equipotential curve

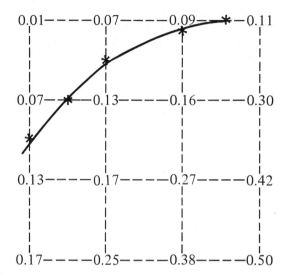

Intensity Scaling

The technique of *intensity scaling* permits the computer to effectively draw the equipotential curves. To illustrate the technique in its simplest form, suppose that it is known that in some region of the xy-plane, the potential V lies in the range $0.00 \leqslant V < 3.00$. Imagine that this region of the plane has been divided using a fine xy-grid, and that at every nodal point we place a single digit JV, given by

$$JV = 0 \text{ for } 0.00 \leqslant V < 1.00,$$
$$JV = 1 \text{ for } 1.00 \leqslant V < 2.00,$$
$$JV = 2 \text{ for } 2.00 \leqslant V < 3.00.$$

In general, the appropriate value of JV at each nodal point can be found using the single FORTRAN statement JV = V, which truncates V to an integer by dropping the decimal part. Since a single digit is used to represent the potential at each nodal point rather than the actual numerical value of the potential, we are able to construct a much finer grid than in version 1. If the digit JV is printed for all nodal points in a grid, we might obtain a result such as that shown in Figure 2.5a. Note that two equipotential curves can immediately be drawn on this figure. The V = 1.00 equipotential is the boundary between the region of 0's and the region of 1's; similarly the V = 2.00 equipotential is the boundary between the 1's and the 2's.

Figure 2.5   Equipotential curves using intensity scaling

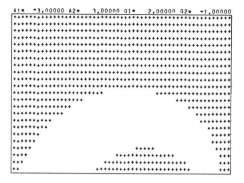

The technique just described can be easily extended to deal with an arbitrary range of the potential. Suppose that we wish to show nine equipotential curves for a series of equally spaced potentials between V_{min} and V_{max}. In this case, a single digit could be obtained for each nodal point using the relation

$$JV = 9.0 \left(\frac{V - V_{min}}{V_{max} - V_{min}} \right). \tag{2.3}$$

Note that the condition $V_{min} \leqslant V \leqslant V_{max}$ restricts JV to the range $0 \leqslant JV \leqslant 9$. As before, we could print the digit JV for all nodal points in the grid to obtain a set of equipotential curves. However, this time we shall modify the procedure to display the same equipotential curves in a more striking way. Instead of printing the digit JV itself, we print a character C, where

      C = ' ' (a blank space), if JV is odd,
      C = '+' (a plus sign), if JV is even and V is nonnegative,
      C = '−' (a minus sign), if JV is even and V is negative.

The borders between regions of plus (or minus) signs and blanks now identify equipotential curves in a more graphic way. (Compare figures 2.5a and 2.5b.)

Equation 2.3 is an example of a *linear scaling function* since JV is a linear function of the potential V. When a linear scaling function is used to compute the digit JV, the equipotential curves obtained correspond to a set of potentials having a constant increment ΔV: $V_{min}, V_{min} + \Delta V, V_{min} + 2\Delta V, \ldots, V_{max}$. Equipotential curves can be found using *any* scaling function for the digit JV, but nonlinear functions do not provide a constant increment between adjacent curves. The *logarithmic scaling function*

$$JV = 2 \log_2 |V| \tag{2.4}$$

has several advantages over the linear scaling function. Figure 2.6 was generated by computer using this logarithmic scaling function and the previously described technique of printing the characters '+', '−', and ' ' instead of the digit JV itself. The absolute value of V appears as the argument of the logarithm in equation 2.4 so that the equation can be used for either sign of the potential V. The logarithm to base two means that JV increases by two if V is doubled. This has two important consequences: (1) Each equipotential curve forming the inner boundary of any region of plus or minus signs represents a potential which is either twice or half as great as that of the equipotential curve forming the next inner boundary. The equipotential curves defined by the inner boundaries of each region of plus or minus signs thus correspond to a series of potentials: $V_{min}, 2V_{min}, 4V_{min}, 8V_{min}, \ldots$ (2) The entire pattern of equipotential curves is not affected by the scale of the problem. If, for example, the potential at every point were doubled (or multiplied by any power of two), the pattern would be entirely unaffected. This property of scale independence is important since it means there is no need to determine values for V_{min} and V_{max}, as in the case of the linear scaling function.

Figure 2.6

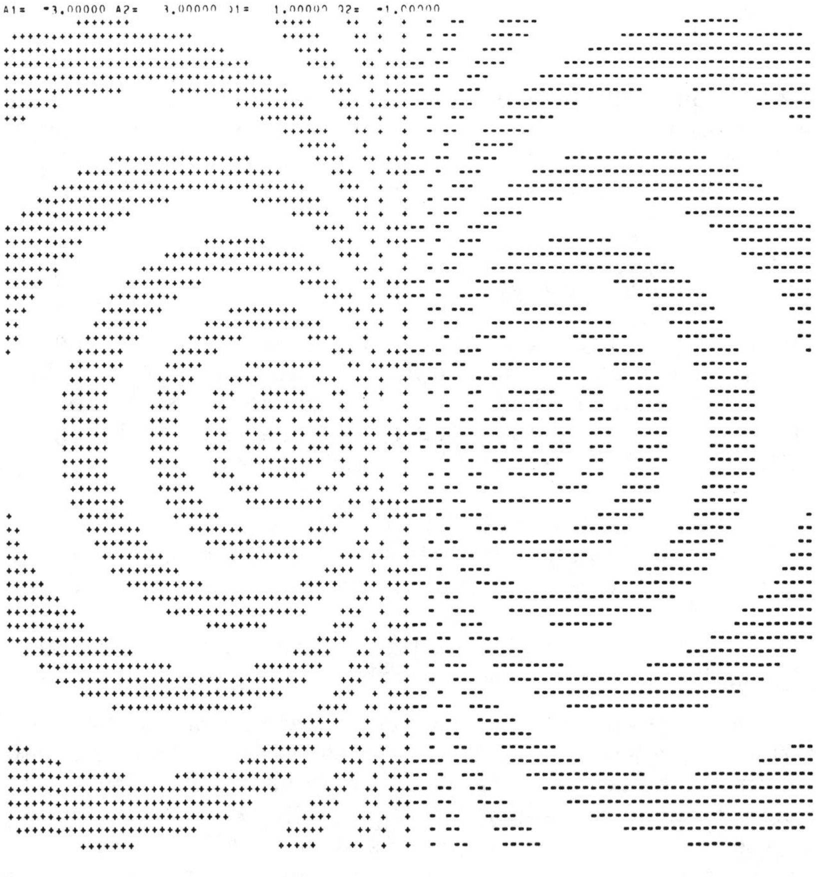

Equipotential Plotting Program (version 2)

The first part of version 2 of the program is essentially the same as version 1. (See flow chart in Figure 2.7, and listing on pages 62-64. In version 2 we use a fine grid of 101 × 61 points instead of the 21 × 21 grid used in version 1. We also use the logarithmic scaling function (equation 2.4) discussed in the previous section. The variable JTEST is the branch which checks whether JV is

Figure 2.7   Diagram for equipotential plotting program (version 2)

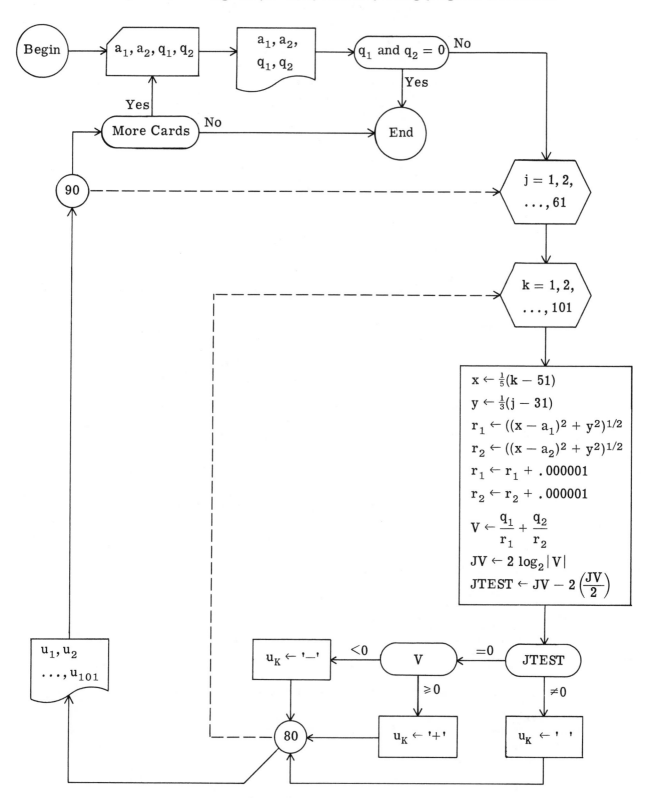

even or odd. The data card used to generate the output shown in Figure 2.8 is the same as the one used with version 1 to generate the output shown in Figure 2.3.

Figure 2.8

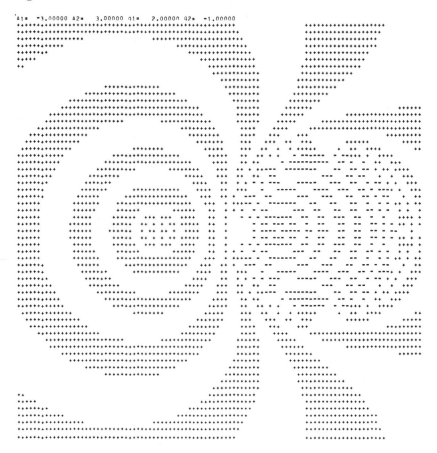

Problems 1 through 9 at the end of this chapter pertain to the Equipotential Plotting Program.

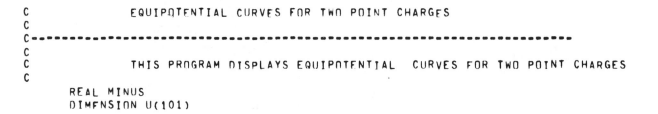

```
C              EQUIPOTENTIAL CURVES FOR TWO POINT CHARGES
C
C---------------------------------------------------------------
C
C          THIS PROGRAM DISPLAYS EQUIPOTENTIAL  CURVES FOR TWO POINT CHARGES
C
       REAL MINUS
       DIMENSION U(101)
```

```
      DATA BLANK,PLUS,MINUS/' ','+','-'/
      ALG2=ALOG(2.0)/2.0
C
C
C              READ A DATA CARD CONTAINING A1, A2, Q1 AND Q2.
C              A1 AND A2 ARE THE X COORDINATES OF CHARGES Q1 AND Q2.
C
   10 READ(2,1000)A1,A2,Q1,Q2
      WRITE(3,1001)A1,A2,Q1,Q2
C
C
C              IF THE DATA CARD READ IN IS BLANK, THEN Q1 AND Q2 WILL BE ZERO.
C
      IF(Q1)30,20,30
   20 IF(Q2)30,100,30
   30 CONTINUE
C
C
C              THIS LOOP GOES OVER 61 ROWS.
C
      DO 90 J=1,61
C
C
C              THIS LOOP GOES OVER 101  COLUMNS WITHIN A ROW.
C
      DO 80 K=1,101
C
C
C              THE X AND Y COORDINATES ARE CALCULATED FROM THE ROW AND COLUMN
C              INDICES J AND K.   RECALL THAT X AND Y MUST LIE BETWEEN -10 AND +10
C
      AJ=J
      AK=K
      X=(AK-51.)/5.
      Y=(AJ-31.)/3.
C
C
C              CALCULATE THE DISTANCES R1 AND R2, FROM THE POINT (X,Y) TO EACH
C              OF THE TWO POINT CHARGES.
C
      R1=SQRT((X-A1)**2+Y**2)
      R2=SQRT((X-A2)**2+Y**2)
C
C
C              THE NEXT TWO STATEMENTS ARE JUST IN CASE R1 OR R2 ARE ZERO
C
      R1=R1+.000001
      R2=R2+.000001
C
C
C              CALCULATE THE NET ELECTRICAL POTENTIAL FOR THE TWO POINT CHARGES.
C
      V=Q1/R1+Q2/R2
C
C
C              THE QUANTITY JV IS SO DEFINED THAT IT INCREASES BY TWO, IF V IS
C              DOUBLED
C
      JV=ALOG(ABS(V))/ALG2
      JTEST=JV-2*(JV/2)
      IF(JTEST)40,50,40
   40 U(K)=BLANK
      GO TO 80
   50 IF(V)70,60,60
   60 U(K)=PLUS
      GO TO 80
```

```
   70   U(K)=MINUS
   80   CONTINUE
   90   WRITE(3,1002)(U(K),K=1,101)
C
C              READ NEXT DATA CARD.
C
         GO TO 10
 1000   FORMAT(4F10.5)
 1001   FORMAT(4H1A1=,F10.5,4H A2=,F10.5,4H Q1=,F10.5,4H Q2=,F10.5)
 1002   FORMAT(1X,101A1)
  100   CONTINUE
         CALL EXIT
         END
```

2.2 Potential for a Charged Thin Wire

In order to calculate the potential for a charged thin wire, we can assume that the net charge on the wire consists of a row of closely spaced point charges. We can easily calculate the potential V at any point (x, y) in a plane containing the wire using equation 2.1. In practice, however, the number of excess electrons on a charged wire is usually so large that we can regard the charge on the wire as continuously distributed along its length, and speak of a *line of charge*. Mathematically, what we have done is to replace the finite sum appearing in equation 2.1 by an integral. Suppose each of the n point charges in the row has a value Δq, such that the total charge of all the n charges is a constant q, independent of n. Then, by definition of a definite integral, the finite sum

$$V = \sum_{j=1}^{n} \frac{\Delta q}{r_j} \tag{2.5}$$

approaches the integral

$$V = \int \frac{dq}{r} \tag{2.6}$$

in the limit as n becomes infinite and Δq tends to zero.

Let us put the integrand in equation 2.6 in a form which can be directly evaluated. We choose the x-axis to lie along the wire, and we assume that the wire extends from $x = -\alpha$ to $x = +\alpha$. The charge dq contained within a given length of wire dx is given by

$$dq = \lambda(x)\,dx, \tag{2.7}$$

where $\lambda(x)$ is some function which gives the charge density at position x. The factor r in equation 2.6 is the distance from a point $(x, 0)$ on the wire to some point (x_0, y_0), at which we wish to find the potential (see Figure 2.9), and r is given by

$$r = ((x - x_0)^2 + y_0^2)^{1/2}. \tag{2.8}$$

Substitution of equations 2.7 and 2.8 into 2.6 yields

$$V = \int_{x=-\alpha}^{x=+\alpha} \frac{\lambda(x)}{((x-x_0)^2 + y_0^2)^{1/2}} \, dx. \tag{2.9}$$

For the case of a wire that has charge uniformly distributed along its length, the function $\lambda(x)$ is simply a constant λ_0. In this case, the integral in equation 2.9 can be directly evaluated, to give

$$V = \lambda_0 \ln \left[\frac{((x_0 - \alpha)^2 + y_0^2)^{1/2} - (x_0 - \alpha)}{((x_0 + \alpha)^2 + y_0^2)^{1/2} - (x_0 + \alpha)} \right]. \tag{2.10}$$

There are only a few special cases (such as $\lambda(x)$ constant) for which the integral in equation 2.9 can be directly evaluated. In other cases, we must use approximate numerical integration techniques, such as those discussed in the next section, to calculate the integral. Knowing the exact solution for $\lambda(x)$ constant serves as a valuable check on the accuracy of the approximate numerical methods.

Numerical Methods for Evaluating a Definite Integral

Let $f(x)$ represent some function that we wish to integrate. For the thin charged wire, $f(x)$ has the form

$$f(x) = \frac{\lambda(x)}{((x-x_0)^2 + y_0^2)^{1/2}}. \tag{2.11}$$

This function is plotted in Figure 2.10 for $\lambda(x) = \lambda_0$ (a constant). Since the value of the integral

$$I = \int_{-\alpha}^{+\alpha} f(x) \, dx \tag{2.12}$$

can be interpreted as the area under the curve between $x = -\alpha$ and $x = +\alpha$, we will divide the area into regions and approximate the areas of the regions. Suppose the value of the function $f(x)$ has been determined at a number of equally spaced x-values $x_1, x_2, \ldots, x_n$, where $x_1 = -\alpha$ and $x_n = +\alpha$. The simplest approximation to the integral is

$$I = \sum_{j=1}^{n} f(x_j) \Delta x, \tag{2.13}$$

where $\Delta x = \dfrac{2\alpha}{(n-1)}$ is the spacing between adjacent x-values. In the specific problem we are interested in, this means using a discrete set of n point charges instead of the continuous line of charge. A comparison of equations 2.5 and 2.13 shows that the individual terms in the sum are the contributions to the potential at (x_0, y_0) due to the charges at $(x_j, 0)$. Figure 2.11 illustrates this for the case n = 5.

The *area* of each of the five rectangles represents the contribution to the potential due to one of five equal point charges $\frac{2}{5}\lambda_0\alpha$ placed at the indicated x-

Figure 2.9   A charged thin wire

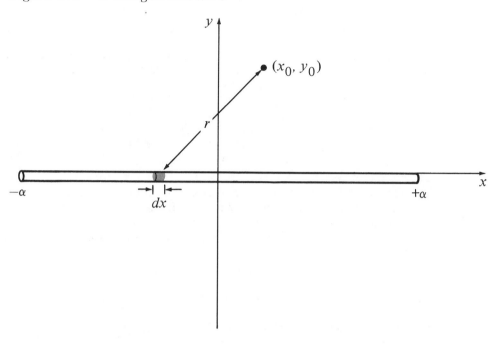

Figure 2.10 $f(\mathbf{x}) = \dfrac{\lambda_0}{((\mathbf{x} - \mathbf{x}_0)^2 + \mathbf{y}_0{}^2)^{1/2}}$

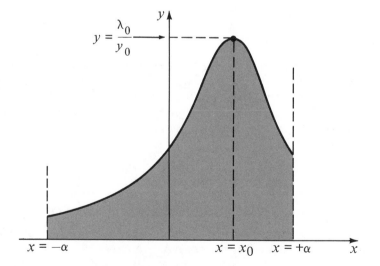

Figure 2.11   Approximation of the integral by a finite sum

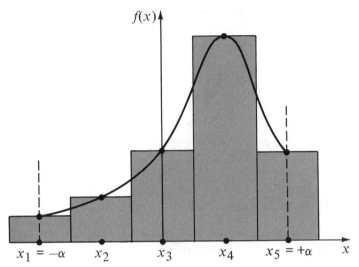

coordinates.* The total area of the five rectangles is a (rough) approximation to the area under the curve between $x = -\alpha$ and $x = +\alpha$.

A better approximation to the integral than equation 2.13 can be obtained if we subtract the excess area which lies outside the interval $-\alpha \leqslant x \leqslant +\alpha$. We subtract one-half the area of the first and last rectangles in Figure 2.11 to obtain

$$I = \sum_{j=1}^{n} f(x_j) \, \Delta x - \tfrac{1}{2} (f(x_1) + f(x_n)) \, \Delta x,$$

which we may write as

$$I = \sum_{j=1}^{n} c_j f(x_j) \, \Delta x, \tag{2.14}$$

where $c_1 = c_n = \tfrac{1}{2}$ and $c_2 = c_3 = c_4 = \ldots = c_{n-1} = 1$. This result is equivalent to the *trapezoidal rule,* in which the area under the curve is approximated by the areas of the four ($n - 1$, in general), trapezoids of Figure 2.12. (Note that equation 2.13 can be considered a special case of equation 2.14 with $c_1 = c_2 = c_3 = \ldots = c_n = 1$.) This raises the question of whether there are any other choices of these coefficients which give an approximation that is better than the trapezoidal rule (equation 2.14). If n is an odd number, we may obtain a significant improvement using *Simpson's rule,* which geometrically corresponds to finding the area under a series of parabolic segments, each passing through three

---

* Note that if we wanted to use a series of five point charges to best approximate a continuous line of charge, we would probably not place them as indicated in Figure 2.11, with two charges at the ends. However, our main concern here is how to best use the values of f(x) at the specified points x_1, $x_2, \ldots, x_n$, to approximate the integral.

Figure 2.12   Trapezoidal approximation to the integral

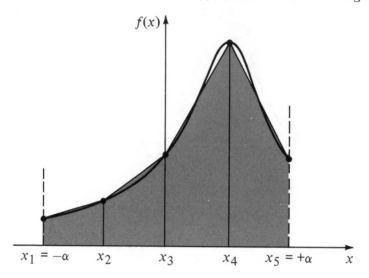

Figure 2.13   Simpson's rule approximation of the integral

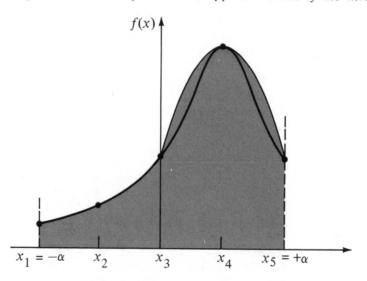

consecutive points of the curve (see Figure 2.13). As we shall shortly prove, the coefficients for Simpson's rule are

c_1	c_2	c_3	c_4	c_5	c_6	$\cdots$	c_{n-1}	c_n
$\frac{1}{3}$	$\frac{4}{3}$	$\frac{2}{3}$	$\frac{4}{3}$	$\frac{2}{3}$	$\frac{4}{3}$	$\cdots$	$\frac{4}{3}$	$\frac{1}{3}$

$$(2.15)$$

(Note that apart from the first and last coefficients, they alternate between $\frac{4}{3}$ and $\frac{2}{3}$.)

In order to derive Simpson's rule, we must first compute the area under the parabola that passes through the three points $(x_1, y_1), (x_2, y_2), (x_3, y_3)$, where $y_i = f(x_i)$. It is surprisingly easy to obtain the equation of this parabola. Consider the function:

$$y(x) = y_1 \underbrace{\left[\frac{(x-x_2)(x-x_3)}{(x_1-x_2)(x_1-x_3)}\right]}_{A} + y_2 \underbrace{\left[\frac{(x-x_1)(x-x_3)}{(x_2-x_1)(x_2-x_3)}\right]}_{B} + y_3 \underbrace{\left[\frac{(x-x_1)(x-x_2)}{(x_3-x_1)(x_3-x_2)}\right]}_{C} \quad (2.16)$$

This function can easily be seen to have the following properties:

1. It is a second order polynomial in x (a parabola).
2. If $x = x_1$, then $A = 1$ and $B = C = 0$, so that $y = y_1$.
3. If $x = x_2$, then $B = 1$ and $A = C = 0$, so that $y = y_2$.
4. If $x = x_3$, then $C = 1$ and $A = B = 0$, so that $y = y_3$.

Hence, equation 2.16 describes a parabola which passes through the three specified points $(x_1, y_1), (x_2, y_2), (x_3, y_3)$.

If equation 2.16 is integrated between $x = x_1$ and $x = x_3$, we can obtain an expression for the area under the parabola which passes through these three points:

$$A_1 = \frac{\Delta x}{3}(y_1 + 4y_2 + y_3),$$

where $\Delta x = x_2 - x_1 = x_3 - x_2$. Similarly, the area under the parabola passing through the three points $(x_3, y_3), (x_4, y_4), (x_5, y_5)$, is given by

$$A_2 = \frac{\Delta x}{3}(y_3 + 4y_4 + y_5)$$

and so on, for any remaining parabolas:*

$$A_3 = \frac{\Delta x}{3}(y_5 + 4y_6 + y_7)$$

$$\vdots$$

$$A_{\frac{n-1}{2}} = \frac{\Delta x}{3}(y_{n-2} + 4y_{n-1} + y_n).$$

---

* In Figure 2.13 there are *only* two parabolas, since there are only five points.

Thus, the total area under all the parabolic segments is given by

$$A = A_1 + A_2 + A_3 + \ldots + A_{\frac{n-1}{2}}$$

or finally

$$A = \frac{\Delta x}{3} (y_1 + 4y_2 + 2y_3 + 4y_4 + \ldots + 4y_{n-1} + y_n),$$

from which Simpson's rule immediately follows.

In deriving Simpson's rule we used equation 2.16, which is the equation of the second order polynomial that passes through the three points (x_1, y_1), (x_2, y_2), (x_3, y_3). It is also surprisingly easy to write down the equation for the lowest order polynomial in x passing through *any* number of points, provided no two points have the same x-coordinate. For example, the third order polynomial which passes through the four points (x_1, y_1), (x_2, y_2), (x_3, y_3), (x_4, y_4), is given by

$$
\begin{aligned}
y = y_1 &\left[\frac{(x - x_2)(x - x_3)(x - x_4)}{(x_1 - x_2)(x_1 - x_3)(x_1 - x_4)} \right] + y_2 \left[\frac{(x - x_1)(x - x_3)(x - x_4)}{(x_2 - x_1)(x_2 - x_3)(x_2 - x_4)} \right] \\
&+ y_3 \left[\frac{(x - x_1)(x - x_2)(x - x_4)}{(x_3 - x_1)(x_3 - x_2)(x_3 - x_4)} \right] + y_4 \left[\frac{(x - x_1)(x - x_2)(x - x_3)}{(x_4 - x_1)(x_4 - x_2)(x_4 - x_3)} \right]
\end{aligned}
\tag{2.17}
$$

(It should be clear from inspection that equation 2.17 is correct—see the discussion on page 69.) Polynomials such as these are usually referred to as *interpolating polynomials*, since they permit a function which has been calculated at a finite set of points (x_1, y_1), (x_2, y_2), $\ldots$, (x_n, y_n), to be approximated by a polynomial of degree $n - 1$, which can then be used to evaluate the function, by interpolation, at any value of x in the range $x_1 \leqslant x \leqslant x_n$.

In the trapezoidal rule, we use *linear* segments, i.e., first order polynomials, to connect adjacent pairs of points on the curve. Similarly, in Simpson's rule, we pass a series of parabolic segments, i.e., second order polynomials, through adjacent triplets of points. Still higher order approximations can be obtained for the integral by passing polynomials of degree m through a number of series of $m + 1$ adjacent points on the curve. This simply involves choosing the appropriate values for the coefficients $c_1, c_2, c_3, \ldots, c_n$, which can then be used with any function f(x). In Table 2.1, we have listed the appropriate coefficients to use, up to the fourth order. As can be seen from their geometrical interpretation, the various methods place restrictions on the values of n (the number of points) that can be used.

It is important to realize that to obtain the increased accuracy which the higher order approximations (usually*) yield, it is only necessary to choose the appro-

---

* The accuracy does not always increase when a higher order approximation is used. In unusual cases a higher order approximation actually may give a poorer result than a lower order approximation, depending on the shape of the function f(x) and the number of terms in the sum n. For example, the trapezoidal rule (Figure 2.12) is *better* than Simpson's rule (Figure 2.13) in the problem of the thin charged wire with $n = 5$.

Table 2.1 Values of coefficients appearing in m^{th} order approximation to the integral: $I = \sum_{j=1}^{n} c_i f(x_i)\, \Delta x$

Order of interpolating polynomials	Name	Minimum n allowed	Other restrictions on n	c_1	c_2	c_3	c_4	c_5	c_6	c_7	c_8	c_9	$\cdots$	c_{n-6}	c_{n-5}	c_{n-4}	c_{n-3}	c_{n-2}	c_{n-1}	c_n
0	finite sum	1	None	1	1	1	1	1	1	1	1	1	$\cdots$	1	1	1	1	1	1	1
1	trapezoidal rule	2	None	$\frac{1}{2}$	1	1	1	1	1	1	1	1	$\cdots$	1	1	1	1	1	1	$\frac{1}{2}$
2	Simpson's rule	3	$n = 2k+1$*	$\frac{1}{3}$	$\frac{4}{3}$	$\frac{2}{3}$	$\frac{4}{3}$	$\frac{2}{3}$	$\frac{4}{3}$	$\frac{2}{3}$	$\frac{4}{3}$	$\frac{2}{3}$	$\cdots$	$\frac{2}{3}$	$\frac{2}{3}$	$\frac{4}{3}$	$\frac{2}{3}$	$\frac{4}{3}$	$\frac{4}{3}$	$\frac{1}{3}$
3	Simpson's $\frac{3}{8}$ rule	4	$n = 3k+1$*	$\frac{3}{8}$	$\frac{9}{8}$	$\frac{9}{8}$	$\frac{3}{4}$	$\frac{9}{8}$	$\frac{9}{8}$	$\frac{3}{4}$	$\frac{9}{8}$	$\frac{9}{8}$	$\cdots$	$\frac{3}{4}$	$\frac{9}{8}$	$\frac{3}{4}$	$\frac{9}{8}$	$\frac{9}{8}$	$\frac{9}{8}$	$\frac{3}{8}$
4	Bode's rule	5	$n = 4k+1$*	$\frac{14}{45}$	$\frac{64}{45}$	$\frac{24}{45}$	$\frac{64}{45}$	$\frac{28}{45}$	$\frac{64}{45}$	$\frac{24}{45}$	$\frac{64}{45}$	$\frac{28}{45}$	$\cdots$	$\frac{24}{45}$	$\frac{28}{45}$	$\frac{64}{45}$	$\frac{24}{45}$	$\frac{64}{45}$	$\frac{64}{45}$	$\frac{14}{45}$

* k = any positive integer

priate values for the coefficients $c_1, c_2, \ldots, c_n$ in equation 2.14. Thus the computer can evaluate an integral using a fourth order approximation in essentially the same amount of time as it takes to evaluate the integral using a first order approximation (for the same value of n). Of course, for a given order m, increased accuracy can also be obtained by using more terms in the sum (larger n). However, the amount of time required for the computer to evaluate equation 2.14 is roughly proportional to n, the number of terms in the sum. Therefore, if higher accuracy is desired, it is usually more efficient to use a higher order approximation instead of more terms in the sum.

Program to Calculate the Potential for a Charged Thin Wire

The program is given in flow diagram form in Figure 2.14, and listed on pages 74-76. The program first reads from a data card numerical values for the parameters N, A, XO, and YO, where

> N = n, the number of point charges, or terms in the sum
> A = α, half the length of the wire, which extends from $-\alpha$ to $+\alpha$
> XO = x_0
> YO = y_0 } coordinates of the point at which potential is to be found.

The program proceeds to calculate the potential according to the exact formula:*

$$V_1 = \lambda_0 \ln \left[\frac{((x_0 - \alpha)^2 + y_0^2)^{1/2} - (x_0 - \alpha)}{((x_0 + \alpha)^2 + y_0^2)^{1/2} - (x_0 + \alpha)} \right]$$

and to calculate the three lowest order approximations: the finite sum (V_2), the trapezoidal rule (V_3), and Simpson's rule (V_4). All three approximations can be written

$$V = \sum_{j=1}^{n} c_j f(x_j) \Delta x,$$

where the values for the coefficients are indicated in Table 2.1 and the function $f(x)$ is given by

$$f(x) = \frac{\lambda(x)}{((x - x_0)^2 + y_0^2)^{1/2}} .$$

In the present version of the program, the function $\lambda(x)$, which is the charge density at point x is assumed to be a constant: $\lambda(x) = 1$ for $-\alpha < x < +\alpha$.

If the program reads the following values of the parameters from data cards:

	N	A	XO	YO
1st card	3.0	1.0	0.0	1.0
2nd card	10.0	1.0	0.0	1.0
3rd card	100.0	1.0	0.0	1.0

---

*   The exact formula can only be used for the case $\lambda(x)$ = a constant.

Figure 2.14   Flow diagram to calculate potential of a charged wire

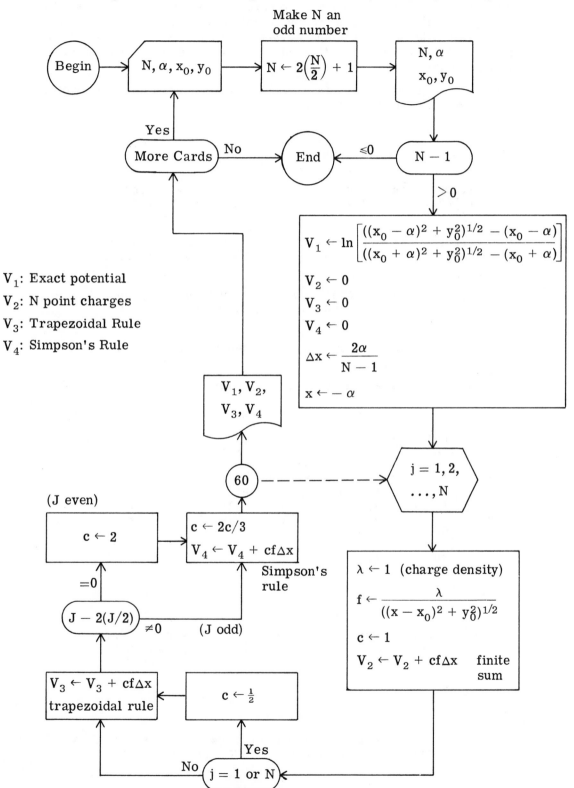

then the output is as shown on Figure 2.15. Note that if N is an even number, as in the case of the second and third cards, the program actually uses the *next higher odd integer*. The reason for this is that Simpson's rule can only be used with odd values of N, and we want to compute $V_1, V_2, V_3,$ and V_4 all for the same value of N, in order to compare the various approximation methods at the same value of N.

Problems 10 through 16 at the end of this chapter pertain to the Potential for a Charged Thin Wire program.

Figure 2.15   Output using sample data

```
      N=    3.00000  A=    1.00000 X0=    .00000 Y0=   1.00000
      V1=   1.76275 V2=    2.41421 V3=   1.70711 V4=   1.80474


      N=   11.00000  A=    1.00000 X0=    .00000 Y0=   1.00000
      V1=   1.76275 V2=    1.90181 V3=   1.76039 V4=   1.76275


      N= 101.00000  A=    1.00000 X0=    .00000 Y0=   1.00000
      V1=   1.76275 V2=    1.77686 V3=   1.76272 V4=   1.76274


      N=    1.00000  A=    .00000 X0=    .00000 Y0=    .00000
```

```
C              ELECTRICAL POTENTIAL FOR A CHARGED WIRE
C
C----------------------------------------------------------------------
C
C              THIS PROGRAM CALCULATES THE ELECTRICAL POTENTIAL FOR A CHARGED
C              WIRE, ASSUMED TO BE A CONTINUOUS FINITE LINE OF CHARGE.
C
C              READ THE PARAMETERS N, A, X0, AND Y0 FROM A DATA CARD.
C              N = THE NUMBER OF POINT CHARGES WHICH MAKE UP THE LINE.
C                  (ALSO THE NUMBER OF TERMS USED TO APPROXIMATE THE INTEGRAL)
C              A = HALF THE LENGTH OF THE LINE, WHICH EXTENDS FROM -A TO +A
C                  ALONG THE X-AXIS
C              X0,Y0 = X AND Y-COORDINATES OF THE POINT AT WHICH THE POTENTIAL
C                  IS TO BE CALCULATED
C
      REAL N,L
      WRITE(3,1003)
   10 READ(2,1000)N,A,X0,Y0
C
C              WE REDEFINE N, MAKING IT THE NEAREST LARGER ODD INTEGER.
C              THIS IS NECESSARY ONLY FOR APPLICATION OF SIMPSONS RULE
C
      M=N
      M=2*(M/2)+1
      N=M
      WRITE(3,1001)N,A,X0,Y0
C
C              IF THE DATA CARD READ IN IS BLANK, THEN N WILL BE ZERO AND WE QUIT
C              (THE REDEFINED VALUE OF N WILL BE ONE, IN THIS CASE.)
```

```
C
      IF(N-1.0)70,70,20
   20 CONTINUE
C
C              THE FOUR POTENTIALS V1, V2, V3, AND V4 ARE DEFINED AS FOLLOWS.
C              V1 = EXACT POTENTIAL FOR A FINITE LINE OF (CONTINUOUS) CHARGE
C              V2 = POTENTIAL FROM SUMMING CONTRIBUTIONS OF N POINT CHARGES
C              V3 = TRAPEZOIDAL RULE APPROXIMATION TO THE LINE OF CHARGE
C              V4 = SIMPSON'S RULE APPROXIMATION TO THE LINE OF CHARGE
C
      V1=ALOG((SQRT((X0-A)**2+Y0**2)-(X0-A))/(SQRT((X0+A)**2+Y0**2)-(X0
     1 +A)))
C
C              V2, V3, V4 ARE COMPUTED AS SUMS OVER THE NUMBER OF CHARGES
C              MAKING UP THE LINE.  SET THEM TO ZERO INITIALLY.
C
      V2=0.
      V3=0.
      V4=0.
C
C          DX IS THE SPACING BETWEEN ADJACENT POINT CHARGES.
C
      DX=2.0*A/(N-1.0)
C
C          X=-A, STARTING AT ONE END OF THE LINE
C
      X=-A
C
C          THIS LOOP IS TO ADD THE POTENTIAL DUE TO EACH OF THE N CHARGES.
C
   DO 60 J=1,M
C
C          L (FOR LAMBDA)  IS THE CHARGE DENSITY AT POSITION X - SET TO 1.
C
      L=1.0
C
C          F IS THE FUNCTION WE WISH TO INTEGRATE.
C
      F=L/SQRT((X-X0)**2+Y0**2)
C
C              FIND C'S FOR FINITE SUM
C
      C=1.0
      V2=V2+C*F*DX
C
C              FIND C'S FOR TRAPEZOIDAL RULE
C
      IF(J-1)30,40,30
   30 IF(J-M)45,40,45
   40 C=0.5
   45 V3=V3+C*F*DX
C
C              FIND C'S FOR SIMPSON'S RULE
C
      IF(J-2*(J/2))55,50,55
   50 C=2.0
   55 C=2.0*C/3.0
      V4=V4+C*F*DX
   60 X=X+DX
```

```
      WRITE(3,1002)V1,V2,V3,V4
C
C              GO READ THE NEXT DATA CARD.
C
      GO TO 10
 1000 FORMAT(4F10.5)
 1001 FORMAT(4H N=,F10.5,4H A=,F10.5,4H X0=,F10.5,4H Y0=,F10.5)
 1002 FORMAT(4H V1=,F10.5,4H V2=,F10.5,4H V3=,F10.5,4H V4=,F10.5//)
 1003 FORMAT(1H1)
   70 CONTINUE
      CALL EXIT
      END
```

2.3   Discharge of a Capacitor in an RC Circuit

If an initially charged capacitor is connected across a resistor, as shown in Figure 2.16, it begins to discharge as soon as the switch S is closed. In this problem we are interested in studying the time dependence of the charge on the capacitor. Once the switch is thrown, the voltage across the resistor V_R must at all times have the same magnitude as the voltage across the capacitor V_C. Since crossing a resistor in the direction of the current represents a potential *drop*, we may write

$$V_R + V_C = 0. \tag{2.18}$$

For the voltage across the resistor, we can make use of Ohm's Law and the definition of electric current to obtain

$$V_R = IR = R\frac{dq}{dt}. \tag{2.19}$$

For the voltage across the capacitor, we use the definition of capacitance to obtain

$$V_C = \frac{q}{C}. \tag{2.20}$$

Upon substituting equations 2.19 and 2.20 into 2.18, we obtain

$$R\frac{dq}{dt} + \frac{q}{C} = 0. \tag{2.21}$$

This first order differential equation may be rearranged to give

$$\frac{dq}{dt} = -\frac{q}{T}, \tag{2.22}$$

where $T = RC$. Rearranging terms and integrating both sides of the equation, we have

$$\int_{q_0}^{q} \frac{dq}{q} = -\frac{1}{T} \int_0^t dt,$$

which yields

$$\ln \frac{q}{q_0} = -\frac{t}{RC}$$

or

$$q = q_0 e^{-t/T},$$

(2.23)

where q_0 represents the charge on the capacitor at time $t = 0$, and T, defined as the product RC, is the "time constant." According to equation 2.23, T is the amount of time needed for q to reach the fraction $1/e \approx 0.368$ of its initial value. In the next section we shall discuss a simple numerical technique, originated by Euler, for obtaining an approximate solution of a first order differential equation such as 2.22. This knowledge of the exact solution shall furnish a valuable check on the accuracy of the approximation.

Figure 2.16   The RC circuit

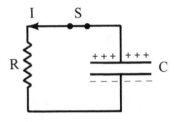

Algorithm Using Euler's Method

Euler's method is a technique which can be used with any first order differential equation once it has been put in the form

$$\frac{dq}{dt} = f(q, t),$$

where $f(q, t)$ is an arbitrary function of q and t. To apply Euler's method we first convert the differential equation (2.22) to a *finite difference equation*, by substituting the ratio of finite quantities $\dfrac{\Delta q}{\Delta t} = \dfrac{q(t + \Delta t) - q(t)}{\Delta t}$ for the derivative dq/dt.

$$\frac{\Delta q}{\Delta t} = -\frac{q}{T}.$$

This can be solved for Δq to yield

$$\Delta q = -\frac{\Delta t}{T} q.$$

(2.24)

Equation 2.24 states that the change in the capacitor's charge, Δq, during the time interval $[t, t + \Delta t]$ is proportional to the amount of charge q on the capacitor at time t. The proportionality constant is the factor $-\Delta t/T$, with the minus sign indicating that the change in charge is negative, that is, the amount of charge decreases with time.

If the initial charge q_0 is known, repeated use of equation 2.24 permits us to find the charge q at a series of times $t = \Delta t, 2\Delta t, 3\Delta t, \dots$. To illustrate the procedure, let us assume the following values: $q_0 = 1$, $T = RC = 10$, and $\Delta t = 1$. We apply 2.24 to find the change in charge during the first time interval Δt:

$$\Delta q = -\tfrac{1}{10} \cdot 1 = -0.1.$$

Thus, the amount of charge remaining at the end of the first time interval is

$$q_1 = q_0 + \Delta q = 0.9.$$

We then use this value of the charge in equation 2.24 to find the change in charge during the next time interval:

$$\Delta q = -\tfrac{1}{10} \cdot 0.9 = -0.09.$$

The amount of charge remaining at the end of the second time interval is

$$q_2 = q_1 + \Delta q = 0.81.$$

By repeating this procedure as many times as desired, we can obtain a numerical solution for the charge q as a function of time. The sequence of values at the end of the first six time intervals is

$$1.0, 0.900, 0.810, 0.729, 0.656, 0.590.$$

These values for $q_0, q_1, \dots, q_5$, may be contrasted with values obtained from the exact solution (equation 2.23) upon substitution of times $t = 0, 1, 2, 3, 4, 5$:

$$1.0, 0.904, 0.818, 0.740, 0.670, 0.606.$$

The agreement is only fair partly because the time step Δt was not chosen small enough (compared to the value of T). However, as we shall see on page 79, the technique can be refined to give better agreement without using smaller values of Δt. This can be of considerable practical importance, since the smaller we make Δt, the more time steps we need to reach a given time, and hence the longer it takes to carry out the computation.

In comparing the above values of the exact and approximate solutions for the first six times, we can infer one other property of the numerical solution of a differential equation, namely, that the error increases with the number of time intervals. The reason for this can be understood by writing the first three terms in the Taylor series expansion for $q(t + \Delta t)$:

$$q(t + \Delta t) = q(t) + \frac{dq}{dt}\Delta t + \tfrac{1}{2}\frac{d^2q}{dt^2}\Delta t^2$$

from which we obtain

$$\frac{\Delta q}{\Delta t} = \frac{q(t + \Delta t) - q(t)}{\Delta t} = \frac{dq}{dt} + \frac{\Delta t}{2}\frac{d^2q}{dt^2} .$$

Thus, to first order in Δt, by using $\Delta q/\Delta t$ in place of dq/dt, we make an error ϵ, given by

$$\epsilon = \frac{\Delta q}{\Delta t} - \frac{dq}{dt} = \frac{\Delta t}{2}\frac{d^2q}{dt^2} .$$

Since the second derivative d^2q/dt^2 of the exponential function $q_0 e^{-t/RC}$ is positive for all values of t, we make an error of the same sign at every time step and the error accumulates.

Improved Euler's Method*

The basis for the Euler method of solving a differential equation is the replacement of the derivative dq/dt at time t by the finite ratio $\Delta q/\Delta t$,

$$\frac{dq}{dt}(t) \approx \frac{\Delta q}{\Delta t} = \frac{q(t + \Delta t) - q(t)}{\Delta t} , \tag{2.25}$$

so that the solution using Euler's method can only be as good as this approximation. An improvement can be made based on the fact that the right side of equation 2.25 is actually a better approximation to the derivative at time $t + \frac{1}{2}\Delta t$ than at time t:

$$\frac{dq}{dt}(t + \tfrac{1}{2}\Delta t) \approx \frac{q(t + \Delta t) - q(t)}{\Delta t} .$$

This can easily be seen from Figure 2.17; the slope of line ⓐ, $\Delta q/\Delta t$, is much closer to that of line ⓒ, $\frac{dq}{dt}(t + \frac{1}{2}\Delta t)$, than to that of line ⓑ, $\frac{dq}{dt}(t)$, independent of the detailed shape of the function q(t). The above equation can be rearranged to yield

$$q(t + \Delta t) \approx q(t) + \frac{dq}{dt}(t + \tfrac{1}{2}\Delta t) \cdot \Delta t. \tag{2.26}$$

In order to use equation 2.26 to find $q(t + \Delta t)$, the value of the charge at the "new" time $t + \Delta t$, we need to know the value of q(t), the charge at the "old" time t, and also the derivative dq/dt at the intermediate time $t + \frac{1}{2}\Delta t$. An approximate value for the derivative at time $t + \frac{1}{2}\Delta t$ can be found by averaging the values at times t and $t + \Delta t$:

$$\frac{dq}{dt}(t + \tfrac{1}{2}\Delta t) \approx \tfrac{1}{2}\left[\frac{dq}{dt}(t + \Delta t) + \frac{dq}{dt}(t)\right] \tag{2.27}$$

---

* This section may be skipped on a first reading, since the RC circuit program discussed in the next section uses only the original, and not the improved, Euler's method.

Figure 2.17   q(t) against t

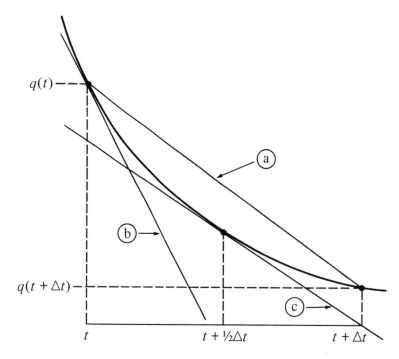

(The derivative at the "average" time $t + \frac{1}{2}\Delta t$ is approximately equal to the average of the derivatives at times t and $t + \frac{1}{2}\Delta t$.)

Substitution of equation **2.27** into **2.26** yields

$$q(t + \Delta t) = q(t) + \frac{\Delta t}{2}\left[\frac{dq}{dt}(t) + \frac{dq}{dt}(t + \Delta t)\right].$$ (2.28)

In order to clarify the meaning of equation **2.28**, we use the subscripts "new" and "old" to signify quantities evaluated at times $t + \Delta t$ and t, respectively. Therefore, equation **2.28** can be rewritten in the form

$$q_{new} = q_{old} + \frac{\Delta t}{2}(q'_{new} + q'_{old}),$$ (2.29)

where the primes indicate derivatives with respect to time.

In Figure 2.18, we see in flow diagram form, the essential steps in the improved Euler's method. Note that to calculate a value for q_{new}, equation 2.29 requires a value for q'_{new}, which itself must be calculated from q_{new}, using equation 2.22! To avoid a circular procedure, we must first calculate an *approximate* value for q_{new} using the original Euler's method. We then use this value to find q'_{new}, with the aid of equation 2.22. Finally, this value of q'_{new} can be used to obtain an *improved* value of q_{new}, using equation 2.29.

The dotted line in Figure 2.18 shows the return path in the loop if the original rather than the improved Euler's method is used.

Figure 2.18   Flow diagram for improved Euler's method

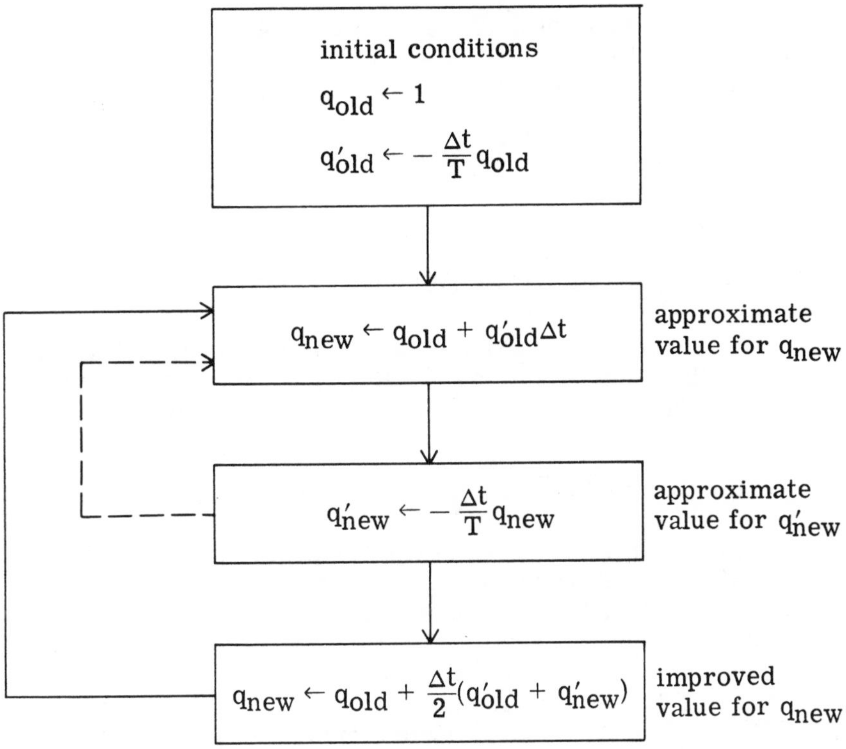

We can illustrate the improved Euler's method using the same values for the parameters used in the illustration of the original Euler's method on page 78: $q_0 = 1, T = 1.0,$ and $t = 0.1$. As in the original Euler's method, we first find an approximate value for q_1 using

$$q_1 = q_0 + q_0' \Delta t.$$

Substituting $T = 1.0$ in the differential equation $q' = -\frac{1}{T} q$ permits us to write $q' = -q$ for any time, so that we find for the approximate value of q_1:

$$q_1 = 1.0 + (-1.0)(0.1) = 0.9.$$

We then find an improved value for q_1 using

$$q_1 = q_0 + \frac{\Delta t}{2} (q_0' + q_1'),$$

where the derivative $q_1' = -q_1$. We therefore obtain

$$q_1 = 1.0 + \tfrac{1}{2} (0.1)(-1.0 - 0.9) = .905.$$

We can repeat the whole procedure to find q_2, first finding an approximate value from

$$q_2 = q_1 + q_1' \Delta t,$$

which yields

$$q_2 = .905 + (-.905)(0.1) = .8145,$$

and then an improved value using

$$q_2 = q_1 + \frac{\Delta t}{2}(q_1' + q_2'),$$

which yields

$$q_2 = .905 + \tfrac{1}{2}(0.1)(-.905 - .8145) = .819.$$

The values for q_1 and q_2 using the improved Euler's method are much closer to the exact values ($q_1 = .904, q_2 = .818$) than the values obtained using the original Euler's method ($q_1 = .900, q_2 = .810$).

RC Circuit Program

The program is given in flow diagram form in Figure 2.19, and listed on pages 85-87. The program first reads from a data card, numerical values for the parameters TAU, DT, and N, where

    TAU = T (time constant)
    DT = Δt (time step)
    N   = number of time steps.

After reading the data card, the program proceeds to advance the time t step by step in the sequence: $0, \Delta t, 2\Delta t, 3\Delta t, \ldots, N\Delta t$. For each value t, the program calculates an exact value for the charge Q1 according to equation 2.23, and an approximate value for the charge Q2 according to Euler's method (the original, not the improved one). The sample output shown in Figure 2.20 was generated using a data card containing the following numerical values for the parameters:

    TAU   DT   N
    10.   1.0   50.

The values of time are listed in the first column.* The exact values of the charge at each time are listed in the second column and plotted as asterisks. The approximate values of the charge at each time are listed in the third column and plotted as plus signs. In addition to printing out the table of numerical values of Q1 and Q2 at each time, the program also displays graphs of Q1 and Q2 as functions of time. To read the graph as q against t, the printout must be turned as in Figure 2.20, to place the origin (0, 0) in the lower left corner.

The graph is produced using a scaling technique that we shall use for all graphs made on the printer. The printer can only display characters at a discrete set of positions on a line, and is therefore not a very accurate plotting device.

---

* Due to truncation which occurs when numbers are printed, the values listed for the time may not be exactly divisible by the time step. For example, 1.49 is printed instead of 1.50 since the number stored in memory is possibly 1.4999999.

Figure 2.19   *Flow diagram for RC circuit program*

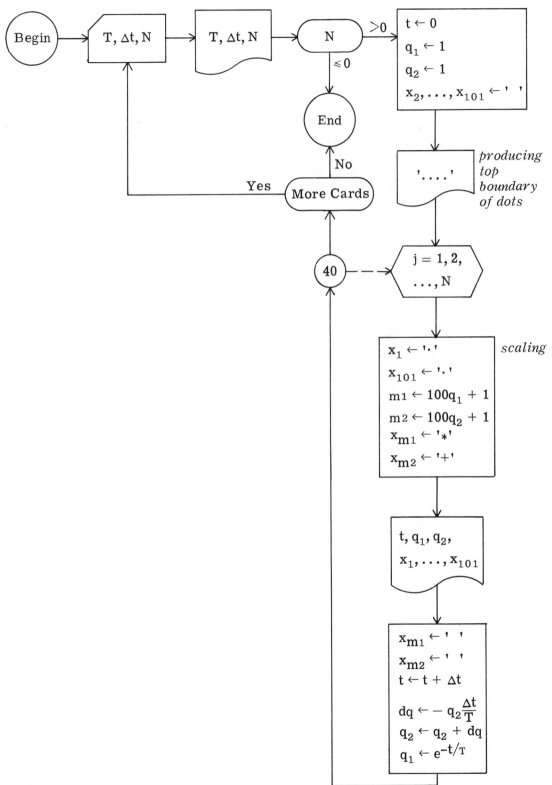

Given variable z which can take on a continuous range of values, we use a scaling function to determine the position on a printed line corresponding to a particular value of the variable z. We normally use a linear scaling function in graphing a function of one variable so as not to distort the plot:

$$M = az + b.$$

The values of the constants a and b are chosen so that the value of M stays within some desired range corresponding to values of z in some specified range. The value of M is truncated using integer arithmetic, so that M can be used as the subscript of an array X. Suppose the array X has 101 elements $X_1, X_2, \ldots, X_{101}$, which can be printed on one line as 101 single characters. Assume the character ' ' (blank) is first stored in every element of the array X. We then use the scaling function to store some other character (for example,

Figure 2.20

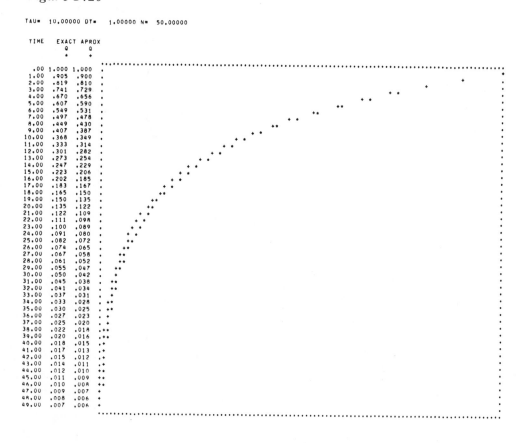

'+' or '*') in a particular element of the array X corresponding to the value of some variable. When the entire array $X_1, X_2, \ldots, X_{101}$ is printed on a line, the position of the (non-blank) character on the line corresponds to the value of the variable. In the RC circuit program, we have two variables (Q1 and Q2) to plot, and we shall use scaling functions of the same form for each:

$$M1 = 100\,Q1 + 1$$
$$M2 = 100\,Q2 + 1$$

(2.30)

The form of these scaling functions insures that M1 and M2 are always within the desired range from 1 to 101. (The smallest values that Q1 and Q2 assume are Q1 = Q2 = 0 (at t = ∞) and the largest values are Q1 = Q2 = 1 (at t = 0).) It is essential that the correct scaling function is used, otherwise M1 and M2 might take on values outside the range from 1 to 101, which can have disastrous consequences when the program is run on the computer.

We can produce the entire graph, with each successive line of characters corresponding to a different value of the time t. Once values of M1 and M2 are computed for a specific value of the time, according to equation 2.30, the program prints out a line of characters which are all* blanks, except for the M1th, which is an asterisk, and the M2th, which is a plus sign. Thus, the positions of the asterisk and the plus sign on the line are linearly dependent on the values of Q1 and Q2, respectively. After it prints a line of characters, the program repeats the procedure for a new value of the time, so that after it prints N successive lines, the graph is complete. The program then reads another data card, if any remain, with a new set of values for the parameters TAU, DT, and N.

Problems 17 through 29 at the end of this chapter pertain to the RC Circuit Program.

```
C               DISCHARGE OF A CAPACITOR IN AN RC CIRCUIT
C
C ================================================================
C
C        THIS PROGRAM CALCULATES AND PLOTS THE AMOUNT OF CHARGE REMAINING
C        ON A DISCHARGING CAPACITOR AS A FUNCTION OF TIME. IT CALCULATES
C        AN APPROXIMATE VALUE FOR THE CHARGE USING AN ITERATIVE PROCEDURE,
C        IN WHICH THE CHANGE IN THE CHARGE DURING A TIME INTERVAL DT, IS
C        GIVEN BY DQ=-Q*DT/TAU. IT ALSO CALCULATES THE EXACT VALUE OF THE
C        CHARGE USING THE EXPONENTIAL FUNCTION  EXP(-T/TAU)
C
C
C        THE DATA STATEMENT DEFINES CHARACTERS NEEDED TO MAKE A PLOT.
      REAL N
C
      DIMENSION X(101)
      DATA BLANK,PLUS,EX,ASTER,DOT/' ','+','X','*','.'/
C
C        THE QUANTITIES TAU, DT, AND N ARE READ FROM A DATA CARD.
C        TAU IS THE TIME CONSTANT, DT IS THE TIME STEP, AND N IS THE NUMBER
```

---

* The first and last characters on the line are not blanks. The character '·' is stored in elements X_1 and X_{101} of the array X to mark the boundaries of the graph.

```
C               OF STEPS DESIRED.
C
   10  READ(2,1000)TAU,DT,N
       WRITE(3,1001)TAU,DT,N
C
C               WE QUIT IF N IS NOT GREATER THAN ZERO.
C
       N1=N
       IF(N)50,50,20
   20  CONTINUE
C
C               WRITE HEADING ON OUTPUT.
C
       WRITE(3,1002)
C
C               Q1 IS THE EXACT VALUE OF THE CHARGE REMAINING.
C               Q2 IS THE APPROXIMATE VALUE OF THE CHARGE REMAINING.
C               AT TIME T=0, BOTH Q1 AND Q2 ARE ASSUMED TO BE 1.0
C
       T=0.
       Q1=1.0
       Q2=1.0
C
C               SET X(1), X(2), X(3),...,X(101) ALL EQUAL TO THE CHARACTER BLANK.
C               THIS IS PRINTED AS A BLANK SPACE ON THE OUTPUT IN 'A' FORMAT.
C               THE VARIABLE X(K) WILL BE PRINTED IN THE K TH POSITION ON A LINE.
C
       DO 30 K=2,101
   30  X(K)=BLANK
       WRITE(3,1004)
       DO 40 J=1,N1
C
C               STORE DOTS IN X(1) AND X(101) FOR BOUNDARY
C
       X(1)=DOT
       X(101)=DOT
C
C               COMPUTE A NUMBER M1, BETWEEN 1 AND 101, TO CORRESPOND TO VALUES
C               OF Q1, WHICH CAN RANGE BETWEEN 0.0 AND 1.0,  AND A NUMBER M2 FOR
C               Q2.
C
       M1=100.*Q1+1.
       M2=100.*Q2+1.
C
C               STORE AN ASTERISK IN X(M1)  TO REPRESENT Q1.
C               STORE A PLUS IN X(M2)  TO REPRESENT Q2.
C
       X(M1)=ASTER
       X(M2)=PLUS
       WRITE(3,1003)T,Q1,Q2,(X(K),K=1,101)
       X(M1)=BLANK
       X(M2)=BLANK
C
C               CALCULATE THE NEXT VALUES OF Q1 AND Q2.
C
       T=T+DT
       DQ=-Q2*DT/TAU
       Q2=Q2+DQ
   40  Q1=EXP(-T/TAU)
       WRITE(3,1004)
C
C               GO READ NEXT DATA CARD.
C
```

```
      GO TO 10
1000 FORMAT(3F10.5)
1001 FORMAT(5H1TAU=,F10.5,4H DT=,F10.5,3H N=,F10.5//)
1002 FORMAT(20H  TIME   EXACT APROX/20H                Q      Q  /
    1 20H            *      +  )
1003 FORMAT(1X,F5.2,2F6.3,2X,101A1)
1004 FORMAT(20X,101H..............................................................
    1.................................................)
 50  CONTINUE
     CALL EXIT
     END
```

2.4 The RLC Series Circuit

In the circuit shown in Figure 2.21, an EMF, or applied voltage V_a, is connected across the series combination of a switch S, a resistor R, an inductor L, and a capacitor C. In this problem we are interested in determining how the current in the circuit varies as a function of time after the switch is closed.

Figure 2.21   The RLC series circuit

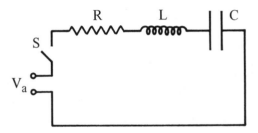

Kirchhoff's second rule, otherwise known as the loop equation, states that around any closed loop, the sum of the potential drops equals the sum of the EMF's. Applying this rule to the circuit shown in Figure 2.21 after the switch is closed gives

$$V_R + V_L + V_C = V_a. \tag{2.31}$$

The voltages across R, L, and C can be expressed in terms of the current I, its time rate of change dI/dt, and the charge on the capacitor q:

$$V_R = IR \qquad V_L = L\frac{dI}{dt} \qquad V_C = q/C. \tag{2.32}$$

Substitution of equations 2.32 into 2.31 yields

$$L\frac{dI}{dt} + IR + q/C = V_a. \tag{2.33}$$

In order to find the current I as a function of time, we must solve equation 2.33 together with the relation

$$I = \frac{dq}{dt},\qquad(2.34)$$

which is the definition of the current flowing away from one plate of the capacitor. This is also the current which flows through the elements R and L because this is a single loop circuit. It should be noted that the two first order differential equations 2.33 and 2.34 are equivalent to a single second order differential equation obtained by differentiating equation 2.33 with respect to time:

$$L \frac{d^2I}{dt^2} + R \frac{dI}{dt} + \frac{I}{C} = \frac{dV_a}{dt}.\qquad(2.35)$$

One interesting special case, which we shall discuss in detail, occurs when the applied voltage V_a has a sinusoidal time dependence:

$$V_a = V_0 \sin \omega t.$$

Note that by choosing $V_0 = 0$, we can include the special case of zero applied voltage. In the next section we shall discuss an algorithm for solving second order differential equations, such as 2.35, which can be used for an arbitrary functional form of the applied voltage V_a.

Algorithm Using Improved Euler's Method

To solve the two first order equations 2.33 and 2.34 (or the one second order equation 2.35), we need two initial conditions. We shall assume that at time $t = 0$, the current I_0 is zero and the charge q_0 is one. With these initial values for I and q, we can use Euler's method, as discussed in the RC circuit problem, to find I and q at a whole series of times. The basis for this iterative procedure is the first order Taylor expansion, from which new values for I and q can be found from the old:

$$\begin{aligned} q_{new} &= q_{old} + q'_{old} \Delta t \\ I_{new} &= I_{old} + I'_{old} \Delta t \end{aligned}\qquad(2.36)$$

and the values of the time derivatives q'_{old} and I'_{old} are found from equations 2.33 and 2.34, using I_{old} and q_{old}.

In order to achieve greater accuracy, we shall use the improved Euler's method. The method is indicated in flow diagram form in Figure 2.22. The dotted line in the figure indicates the return path in the loop for the ordinary Euler's method, in which case "improved" values of the charge and current are not calculated. The basis for the improved Euler's method, like the original method, is also a first order Taylor expansion. However, for the improved method, the derivatives q'_{old} and I'_{old}, appearing in equation 2.36 are replaced by the average values: $\frac{1}{2}(q'_{old} + q'_{new})$, and $\frac{1}{2}(I'_{old} + I'_{new})$. (For a justification of this step, see page 79.)

As indicated in Figure 2.22, the necessary sequence of steps in the improved Euler's method is as follows. First, calculate approximate values for q_{new} and I_{new}, as in the ordinary Euler's method. Then, use these approximate values to obtain approximate values for the derivatives q'_{new} and I'_{new}. For q'_{new}, we use equation 2.34:

$$q'_{new} = I_{new}$$

and for I'_{new}, we solve equation 2.33, to obtain

$$I'_{new} = \frac{1}{L}\left(V_a - I_{new}R - \frac{q_{new}}{C}\right).$$

Finally, calculate improved values for q_{new} and I_{new}, using equation 2.29 for q and an analogous equation for I.

Figure 2.22   Flow diagram for improved Euler's method

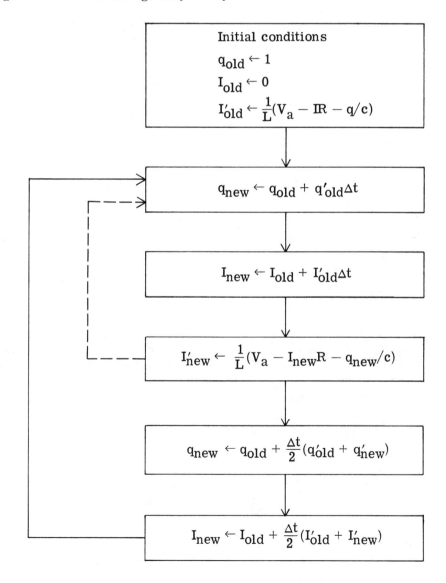

Free Oscillations

If the applied voltage V_a is zero, then in certain cases free oscillations can occur in the RLC series circuit. If V_a is taken to be zero for all times, then dV_a/dt is also zero, in which case equation 2.35 has the exact solution:

$$I = I_0 e^{-(R/2L)t} \sin \omega_0 t, \qquad (2.37)$$

where

$$\omega_0 = \frac{1}{\sqrt{LC}} \left(1 - \frac{R^2 C}{4L}\right)^{1/2}. \qquad (2.38)$$

The computer-generated curve of Figure 2.23 is the solution obtained using the improved Euler's method. The gradually decaying oscillations of the curve agree with the general form of the exact solution, in which an oscillatory factor, $\sin \omega_0 t$, is multiplied by a "damping factor" $e^{-(R/2L)t}$. As long as $R^2 C/4L$ is much less than one, then equation 2.38 which gives a value for ω_0, reduces to

$$\omega_0 \approx \frac{1}{\sqrt{LC}} \qquad \left(\text{for } R \ll \sqrt{\frac{4L}{C}}\right); \qquad (2.39)$$

ω_0 is the frequency of the free oscillations in radians per second. The period T of the oscillations is given by

$$T = \frac{2\pi}{\omega_0} \approx 2\pi \sqrt{LC}. \qquad (2.40)$$

The appearance of the damping factor $e^{(-R/2L)t}$, in equation 2.37 implies that the *envelope*, i.e., the non-oscillatory curve drawn tangent to the solution (see the hand-drawn dotted curve in Figure 2.23) should fall to a fraction $1/e \approx .368$ of its initial value after a time

$$\tau = \frac{2L}{R}. \qquad (2.41)$$

By examining the expression for ω_0 (equation 2.38) we see that the condition for oscillations to occur is that R should be small compared to $\sqrt{4L/C}$. In order that ω_0 be a real number, we must have

$$1 - \frac{R^2 C}{4L} > 0,$$

and therefore

$$R < \sqrt{\frac{4L}{C}}. \qquad (2.42)$$

If R exceeds the critical value $\sqrt{4L/C}$, then ω_0 becomes imaginary. In this case, the factor $\sin \omega_0 t$ is actually equivalent to the exponential of a real number, and therefore the solution becomes a pure exponential and no longer oscillates. For this reason, when R has the value

$$R = \sqrt{\frac{4L}{C}}$$

the circuit is said to be *critically damped*.

Figure 2.23

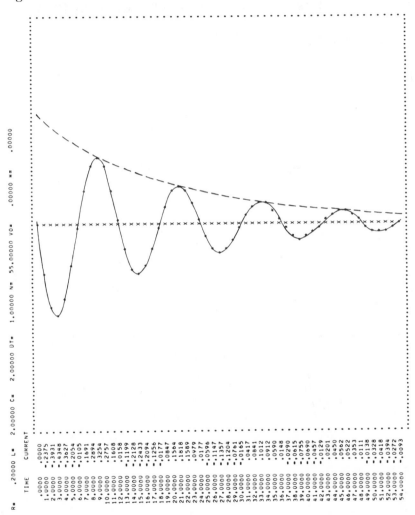

Forced Oscillations

If the applied voltage is a sinusoidal function of time, then forced oscillations occur in the RLC circuit. When the applied voltage is given by $V_a = V_0 \sin \omega t$, then equation 2.35 again has an exact solution. If the initial conditions are chosen properly, the current is a sinusoidal function of time, with the same oscillation frequency as the applied voltage:

$$I = I_m \sin(\omega t - \phi),$$

(2.43)

where

$$I_m = \frac{V_0}{Z},$$

(2.44)

and the quantity Z is known as the *impedance* and is given by

$$Z = \left(R^2 + \left(\omega L - \frac{1}{\omega C}\right)^2\right)^{1/2}.$$

(2.45)

Figure 2.24

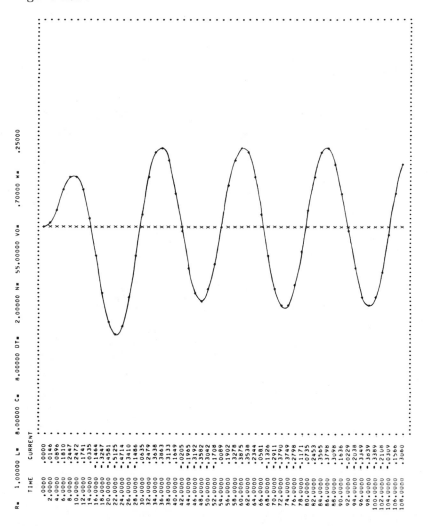

The quantity ϕ appearing in equation 2.43 is the *phase shift* which is the relative shift between the sine curves describing the applied voltage and the current as functions of time; ϕ is given by

$$\phi = \tan^{-1} \frac{\left(\omega L - \frac{1}{\omega c}\right)}{R} . \qquad (2.46)$$

If the initial conditions are chosen arbitrarily, the solution will not have a purely sinusoidal time dependence. In Figure 2.24, we see a computer-generated solution obtained using the improved Euler's method with the initial conditions $q_0 = 1$, $I_0 = 0$. The curve looks nonsinusoidal at first, but then seems to assume a sinusoidal shape. Such initial departures from a sinusoidal shape are known as *transient currents,* and they occur whenever the initial conditions are other than those consistent with the specific solution which is purely sinusoidal (equation 2.43). For all solutions, once the solution becomes sinusoidal, the amplitude of the curve and its oscillation frequency depend only on the values of the parameters R, L, C, and ω, and not on the initial conditions.

Like any system whose oscillations are driven by an external source, an RLC circuit can resonate when the oscillation frequency of the external source matches the "natural" frequency of the circuit. (The natural frequency is the frequency of free oscillations.) At resonance the amplitude of the current has a greater value than for any other frequency. To find the frequency at which the amplitude of the current $I_m = \frac{V_0}{Z}$ reaches its maximum, we must find the frequency for which Z is a minimum. Since the factor $\left(\omega L - \frac{1}{\omega C}\right)^2$ which appears in equation 2.45 can not be negative, Z will have its smallest possible value when this factor is zero:

$$\left(\omega L - \frac{1}{\omega C}\right)^2 = 0 \qquad (2.47)$$

Solving equation 2.47 for ω yields

$$\omega = \frac{1}{\sqrt{LC}} . \qquad (2.48)$$

Note that the resonant frequency given by equation 2.48 is the same as the frequency of free oscillations, given by equation 2.39. This means that the circuit shows the greatest "response" (maximum amplitude current), when the applied voltage has a frequency equal to the natural frequency, i.e., the frequency of free oscillations.

RLC Circuit Program

The program is given in flow diagram form in Figure 2.25, and listed on pages 95-96. The program first reads from a data card numerical values for the parameters R, L, C, DT, N, VO, and W, where

    R = resistance
    L = inductance
    C = capacitance

Figure 2.25   Flow diagram for RLC circuit program

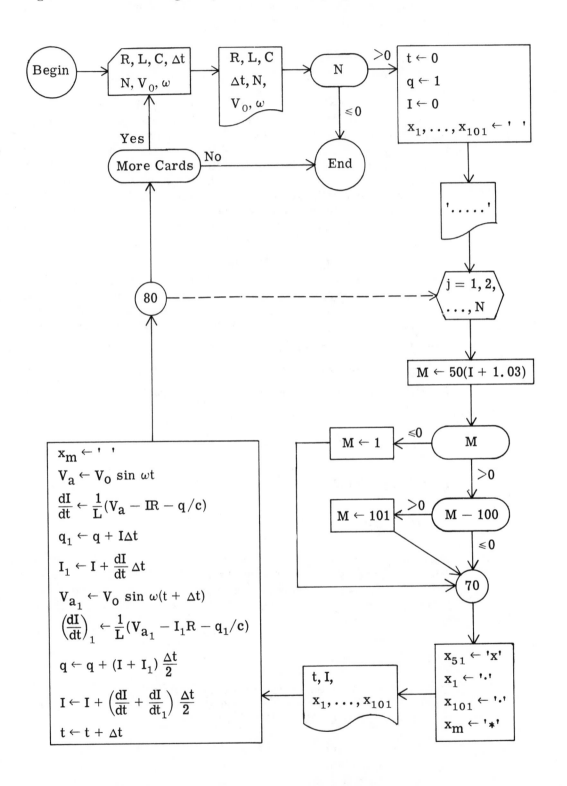

$DT = \Delta t$ (time step)
N = number of time steps
$VO = V_0$ (amplitude of applied voltage)
$W = \omega$ (frequency of applied voltage).

After reading the data card and setting the initial conditions, the program proceeds to advance the time in the sequence $0, \Delta t, 2\Delta t, 3\Delta t, \ldots, N\Delta t$. For each new value of the time, it calculates new values of the charge q_{new}, and current I_{new} from the old values, using the improved Euler's method. In addition to printing out a table of numerical values giving the current at each time, the program also displays a graph of current against time, using a technique described in the RC circuit problem.

```
C            RLC SERIES CIRCUIT
C
C.........................................................................
C
C            THIS PROGRAM CALCULATES AND PLOTS THE CURRENT AS A FUNCTION OF
C            TIME FOR AN RLC SERIES CIRCUIT IN WHICH THE CAPACITOR IS INITIALLY
C            CHARGED  AND THE CURRENT IS INITIALLY ZERO.
C
C            THE DATA STATEMENT DEFINES CHARACTERS TO BE USED IN MAKING PLOTS.
C
      DATA BLANK,EX,ASTER,DOT/' ','X','*','.'/
      REAL L,I,I1,N
      DIMENSION X(101)
C
C            READ A DATA CARD CONTAINING R, L, C, DT, N, VO, AND W
C            R=RESISTANCE, L=INDUCTANCE, C=CAPACITANCE, DT=TIME STEP,
C            VO=AMPLITUDE OF APPLIED VOLTAGE, W=FREQUENCY OF APPLIED VOLTAGE,
C            N=NUMBER OF TIME STEPS
C
   10 READ(2,1000)R,L,C,DT,N,VO,W
      WRITE(3,1001)R,L,C,DT,N,VO,W
C
C            WE QUIT IF N IS NOT GREATER THAN ZERO.
C
      N1=N
      IF(N)90,90,20
   20 CONTINUE
C
C            THE ASSUMED INITIAL CONDITIONS ARE I=0 AND Q=1 AT T=0.
C
      T=0.
      Q=1.0
      I=0.0
C
C            SET X(1), X(2), X(3),.... X(101) ALL EQUAL TO THE CHARACTER BLANK.
C
      DO 30 K=1,101
   30 X(K)=BLANK
      WRITE(3,1003)
      DO 80 J=1,N1
C
C            COMPUTE AN INDEX M, IN THE RANGE 0 TO 100 TO CORRESPOND TO A
C            VALUE OF THE CURRENT IN THE RANGE -1.0 TO +1.0
C
```

```
      M=50.*(I+1.03)
C
C              THE NEXT 5 STATEMENTS RESTRICT M TO THE INTERVAL 1 - 101
C
      IF(M)40,40,50
   40 M=1
   50 IF(M-100)70,70,60
   60 M=101
   70 CONTINUE
C
C              STORE THE CHARACTER 'X' IN X(51), THE 51ST  POSITION ON A LINE,
C              TO REPRESENT THE X-AXIS.
C
      X(51)=EX
C
C              STORE THE CHARACTER '*' IN X(M).
C
      X(1)=DOT
      X(101)=DOT
      X(M)=ASTER
C
C              WRITE OUT THE RESULTS FOR TIME T, AND STEP THE TIME BY DT.
C
      WRITE(3,1002)T,I,(X(K),K=1,101)
      X(M)=BLANK
C
C              VA IS THE APPLIED VOLTAGE SIGNAL.
      VA=V0*SIN(W*T)
C
C              DIDT IS THE TIME DERIVATIVE OF I.
C
      DIDT=(1.0/L)*(VA-I*R-Q/C)
C
C              FIRST FIND APPROXIMATE VALUES FOR Q AND I FROM THE OLD ONES.
C
      Q1=Q+I*DT
      I1=I+DIDT*DT
      VA1=V0*SIN(W*(T+DT))
      DIDT1=(1.0/L)*(VA1-I1*R-Q1/C)
C
C              THEN GET IMPROVED NEW VALUES FOR Q AND I
C
      Q=Q+(I+I1)*DT/2.0
      I=I+(DIDT+DIDT1)*DT/2.0
      T=T+DT
   80 CONTINUE
      WRITE(3,1003)
C
C              READ NEXT DATA CARD IF ANY.
C
      GO TO 10
 1000 FORMAT(7F10.5)
 1001 FORMAT(3H1R=,F10.5,3H L=,F10.5,3H C=,F10.5,
     1  4H DT=,F10.5,3H N=,F10.5,4H V0=,F10.5,3H W=,F10.5//
     2 20H      TIME     CURRENT)
 1002 FORMAT(1X,2F8.4,2X,101A1)
 1003 FORMAT(19X,101H...................................................
     1...................................................)
   90 CONTINUE
      CALL EXIT
      END
```

Discussion of Sample Results

The sample output shown in Figures 2.23 and 2.24 were obtained using two data cards with the following numerical values for the parameters:

	R	L	C	DT	$\dot{N}$	VO	W
Figure 2.23	0.2	2.0	2.0	1.0	55.0	0.0	0.0
Figure 2.24	1.0	8.0	8.0	2.0	55.0	0.7	0.25

In the first case, since the applied voltage is zero (VO = 0), and since the values used for R, L, and C satisfy inequality 2.42, the solution is a damped oscillatory one, as previously discussed. In the second case, the sinusoidally varying applied voltage (VO $\neq$ 0) yields a solution which is also sinusoidal, apart from a transient effect.

Problems for Chapter 2

Equipotential Plotting Program

1. Equipotential curves and electric field lines.   Run versions 1 and 2 of the program using a number of data cards and discuss the results. Among the values you select for the charges Q1 and Q2, you might try pairs for which the ratio Q1/Q2 = +1, −1, and −2. For each of these cases, you can try several different values for A1 and A2. For version 1, it is necessary to draw equipotential curves by hand using the interpolation procedure discussed on pages 57-58. For version 2, the computer effectively draws the equipotential curves. In either case, once the equipotential curves are drawn, these can be used to determine the electric field lines. Electric field lines start on positive charges and end on negative charges, or else go off to infinity, and they are everywhere perpendicular to the equipotential curves.

2. The dipole.   A dipole consists of a pair of equal and opposite charges of magnitude q, separated by a distance d. The *dipole moment* p is defined as

$$p = qd.$$

Observe and discuss the equipotential plots and electric field lines for a number of dipoles having the same dipole moment, say p = 6. Use a series of data cards on which a_1 and a_2 take on progressively smaller values, and q_1 and q_2 take on progressively larger values, ranging from ±1 to ±10^8. (The values used for each case should satisfy $a_1 = -a_2$ and $q_1 = -q_2$.)

3. More than two point charges.   Modify version 2 of the program so that it will generate equipotential curves for the case of more than two point charges located at arbitrary coordinates read from a data card. Discuss the results you obtain when you run the program for several cases. For example, you might try four point charges on the corners of a square, or a row of five equally spaced point charges of equal magnitude.

4. The line of charge.   A charged thin wire may be considered a continuous line segment of charge. If the charge is uniformly distributed along the length of the wire, the potential at any point (x_0, y_0) in a plane containing the wire is given by equation 2.10. The factor λ_0 in this equation is the charge per unit length. The factor α is half the length of the wire, which is assumed to extend from $x = -\alpha$ to $x = +\alpha$ along the x-axis.

Modify the program to read values for λ_0 and α from a data card, and then to compute the potential using equation 2.10 for the finite line of charge instead of equation 2.2 for two point charges. Try running the modified version of the program using various values for λ_0 and α, and discuss the results. In particular, compare the equipotential plot for the line of charge with that of five point charges in a row (see problem 3). To make the comparison, use a value for the length of the line which is equal to the distance between the first and fifth charges in the row. Also, use a value for the charge density which is equal to the total charge of all five point charges, divided by the length of the line.

5. *Equally spaced potentials.*   Modify version 2 of the program so that the equipotential curves correspond to a set of equally spaced potentials: V_0, $V_0 + \Delta V, V_0 + 2\Delta V, \ldots$, rather than the geometric sequence $V_0, 2V_0, 4V_0, \ldots$, as assumed in the present version. Discuss the relative merits of the two types of potential spacing, based on your observations of the output obtained using the modified and original versions with several sets of data.

6. *The circular ring of charge—uniform distribution.*   A circular ring of radius R lying in the xy-plane contains a charge q which is uniformly distributed around the ring. This situation may be approximated by N uniformly spaced point charges, each of magnitude q/N, which lie on a circle of radius R and at angles $\theta_n = 2\pi n/N$, where $n = 1, 2, \ldots, N$.

Modify version 2 of the program so that it reads values for R, q, and N from a data card, calculates the coordinates (x_n, y_n) for each of the N charges on the ring, and uses these to create an equipotential plot. Run the modified version of the program using a range of values for N, with q and R fixed at suitable values. For large values of N we would expect that outside the ring the equipotential curves would be nearly concentric circles, whereas inside the ring the entire region should be approximately one equipotential surface.

7. *The circular ring of charge—alternating signs.*   Modify version 2 of the program so that it generates equipotential plots for a ring of N uniformly spaced charges as described in problem 6. However, this time make the point charges of the ring alternate in sign. (For even values of N, the total charge on the ring is therefore zero.) Using the modified version of the program, generate equipotential plots for N = 4, 8, 16, 32. The case of N = 2 would correspond to a pair of equal and opposite charges (a dipole). By extension, N = 4 corresponds to a "quadrupole," N = 8 to an "octapole," etc. Collectively, charge distributions of the type we have described are known as "multipoles." It can be shown that for the Nth multipole, the potential varies with distance from the center of the ring r according to

$$V \approx r^{-(1 + \log_2 N)}, \tag{2.49}$$

for r much greater than R. Thus, at large distances from a dipole, the potential varies as r^{-3}; at large distances from a quadrupole it varies as r^{-4}; at large distances from an octapole it varies as r^{-5}, etc. Use the computer-generated equipotential plots for the various multipoles to verify the predicted r dependence of the potential in each case. (Be sure to use a ring of small radius, since equation 2.49 holds only for $r \gg R$.)

8. The circular ring of charge—a nonuniform distribution.   Modify version 2 of the program so that it generates equipotential plots for a ring of N uniformly spaced charges as described in problem 6. However, this time allow the magnitude of each charge to be a function of its angle θ.

Run the modified version of the program using a large value for N (perhaps N = 50), and try several functions q = q(θ) which specify the charge as a function of angle. See if you can find, by trial and error, that function for which the equipotential curves inside the ring are a series of parallel straight lines (a uniform electric field).

9. Two parallel lines of charge.   Modify version 2 of the program to generate equipotential plots for two charged thin wires (see problem 4). Assume the two wires to be parallel, of equal length, and to contain equal and opposite charges. Run the modified version of the program using a range of values for the separation between the wires while keeping their length fixed.

Potential for a Charged Thin Wire

10. Dependence of the potential on N and YO.   Run the program using a number of data cards on which the values of the parameters A, XO, and YO are fixed, while N takes on a range of values. Repeat this using other values for YO. Finally, try keeping A, XO, and N fixed at some values, while YO takes on a range of values; and repeat this using other values for N. In selecting values for N, bear in mind that the amount of computer time needed is roughly proportional to N.

Discuss the rates at which the three approximate values V_2, V_3, and V_4 converge on the exact value V_1, as N is increased. Also, discuss the dependence of V_1, V_2, V_3, and V_4 on YO for each fixed value of N. In particular, comment on the dependence of V_2 on YO for values of YO much greater than A, and values much less than A/N.

11. Dependence of the potential on XO.   Modify the program to calculate and print V2 (the potential due to N point charges) at a series of x values $x_0, x_0 + dx$, $x_0 + 2dx, \ldots, x_0 + M dx$. Run the modified version of the program using A = 100, N = 200, $x_0 = 0$, dx = 0.2, M = 50, and a range of values for y_0: 0.5, 1.0, 2.0, 4.0. For each value of y_0 make a plot of V2 versus x_0 and discuss the results.

12. Nonuniform charge distribution.   Modify the program so that it calculates the potential due to a charged wire having a nonuniform charge density. Try several functions for the charge density as a function of position along the wire: $\lambda(x)$. One interesting possibility is $\lambda(x) = \sin k\pi x$, which changes sign k − 1 times along the length of the wire. Run the modified version of the program for various values of k and the other parameters, and discuss the same points raised in question 10. (Note that while the three approximate values for the potential V2, V3, and V4 can be found as before, the exact value of the potential V1 can no longer be calculated.)

13. Third and fourth order approximations.   Modify the program so that it calculates the potential due to a charged wire using the third and fourth order approximations for the integral. See Table 2.1 for the appropriate values of the coefficients $c_1, c_2, c_3, \ldots, c_n$, to use in equation 2.14. Note that only certain values of n may be used (see Table 2.1). Run the modified version of the program using various values of N (=n), and compare the various order approximations with the exact potential V_1 in each case.

14. *Amount of computer time.*   Determine the relative amounts of computer time to evaluate the potential for each order approximation (zero through fourth), for various values of n. (To do this, calculate only one potential each time, not four as is the case in the original version of the program.) Make graphs of the error as a function of computer time for each order approximation. (The error can be easily determined, since the exact potential V_1 is calculated as well as the approximate values.)

15. *Interpolation polynomials.*   Write a program to use the second order interpolation polynomial (see equation 2.16) to approximately evaluate a function $y = f(x)$ at any value of x in the interval $x_1 \leqslant x \leqslant x_3$, given numerical values for the coordinates of the three points (x_1, y_1), (x_2, y_2), (x_3, y_3). To test the program, use three points that lie exactly on a sine curve, and use equation 2.16 to find values for sin x, by interpolation, for various values of x. Repeat the procedure using other triplets of points.

16. *Integration of a numerical function.*   Let $y = f(x)$ designate a function that is specified only at a discrete set of points (x_j, y_j), $j = 1, 2, \ldots, n$. Write a program that reads a series of data cards containing numerical values for these n coordinate pairs and calculates the integral of $f(x)$ between $x = x_1$ and $x = x_n$ using the trapezoidal rule and Simpson's rule. Run the program using numbers from a table of sines or cosines as data, and compare your results with the expected values.

RC Circuit Program

17. *Agreement between the exact and approximate solutions.*   Run the program using a number of values for TAU, DT, and N, and discuss how the agreement between the exact and approximate solutions depends on the values of the parameters.

18. *Charging a capacitor in an RC circuit.*   If we apply Kirchhoff's loop equation to the case of a capacitor in an RC circuit being charged by a battery of EMF E_0, we obtain

$$IR + q/C = E_0 .$$

Using $I = \dfrac{dq}{dt}$, this can be rearranged to yield

$$\frac{dq}{dt} = E_0/R - q/RC. \tag{2.50}$$

As can be verified by direct substitution, the exact solution of this equation is

$$q = q_\infty (1 - e^{-t/RC}),$$

Where the charge on the capacitor after an infinite amount of time, q_∞, is given by $q_\infty = CE_0$. The initial condition assumed is that the charge on the capacitor is zero at time $t = 0$.

Equation 2.50 can also be solved approximately using Euler's method by making the replacement $\frac{\Delta q}{\Delta t}$ for $\frac{dq}{dt}$, so that we obtain

$$\frac{\Delta q}{\Delta t} = E_0/R - q/RC$$

or

$$\Delta q = (E_0/R - q/RC)\Delta t, \qquad\qquad (2.51)$$

giving finally

$$\Delta q = \frac{(q_\infty - q)}{RC}\,\Delta t. \qquad\qquad (2.52)$$

We can then repeatedly use equation 2.52 to find the change in charge at a series of times assuming the initial charge to be zero.

Modify the program to calculate and plot exact and approximate values of the charge on a capacitor as a function of time, for the case of a charging capacitor. Run the modified version of the program using $q_\infty = 1.0$, with various values for the parameters TAU (= RC), DT, and N.

19. RC circuit with a square wave input.   If a voltage source which produces a time-dependent voltage $E_0(t)$ is connected across an RC combination, equation 2.51 can still be used to find an approximate solution of 2.50 using Euler's method. (In general, the exact solution can no longer be found.)

Modify the program to calculate and plot the approximate solution using several different functions for the source voltage as a function of time. One interesting case is the square wave, which can be defined by

$E_0 = +V_0$ (a constant), for $\sin t \geqslant 0$

$E_0 = -V_0$, for $\sin t < 0$.

(This is just one of many ways to define the square wave which is reasonably convenient to translate into computer instructions. An even simpler definition, in terms of translation into a single FORTRAN instruction, is

$$E_0 = V_0 \frac{\sin t}{|\sin t| + 10^{-6}}\,.$$

The factor of 10^{-6} in the denominator makes it unnecessary to test for the indeterminate form $0/0$ which would otherwise appear for $t = n\pi$.)

Run the modified version of the program using a square wave for the function $E_0(t)$ with several different values for the parameters TAU, DT, and N. Try using values of TAU much larger than 1.0, equal to 1.0, and much smaller than 1.0. Discuss the shape of the approximate solution in each of these cases. (Do not bother to calculate and plot the exact solution.)

20. Improved Euler's method.   Modify the program so that it uses the improved Euler's method discussed in section 2.3. Run the modified version of the program using various values for the parameters TAU, DT, and N, and discuss the relative agreement of the original and improved methods with the exact solution.

Determine, by trial and error, how much the time step DT must be decreased so that the original and improved Euler's methods have the same accuracy. Compare the relative amounts of computer time required in these two cases—is it more efficient to use the improved Euler's method or smaller time steps?

RLC Circuit Program

21. Free oscillations.   Run the program using a number of data cards with VO = 0 and various values for the other parameters. Verify that the inequality 2.42 must be satisfied if the solution is to be an oscillatory one, by using values of R slightly less than and slightly greater than the value for critical damping. For the oscillatory solutions, see if their detailed shape agrees with the form of the exact solution (equation 2.37). In particular, check that the period (see equation 2.40) and the *damping time* (see equation 2.41) agree with the predicted values.

22. Forced oscillations.   Run the program using a number of data cards with various values for the parameters, except for VO which is kept at some fixed nonzero value. In particular, try a range of values for W with the other parameters held fixed. See if the maximum amplitude current is observed at the predicted resonant frequency, (equation 2.48). Use enough values for W, above and below resonance, to be able to plot a resonance curve (amplitude of the current versus frequency).

23. Other initial conditions.   Modify the program so that the calculation uses initial conditions other than $q_0 = 1, I_0 = 0$. The best way to do this would be to have these two additional parameters read from the data card. Run the program using various initial conditions, but keep the other parameters fixed, and observe the effect on the transient currents.

24. Goodness of the approximation.   Modify the program so that it calculates the exact solution (equation 2.37) and the approximate solution using the ordinary Euler's method, as well as the improved Euler's method solution. Make the program display the three curves on the same graph, using different symbols for each one. Discuss the differences you observe between the exact solution and the two approximate solutions for various choices of the parameters. You should use values of the parameters which include both the cases of many time steps per oscillation and few time steps per oscillation. (In order for the exact and approximate solutions to agree, the constant I_0, appearing in equation 2.37, must be 2/RC.)

25. Nonsinusoidally varying applied voltage.   Modify the program so that the applied voltage is some nonsinusoidal function of time. One interesting possibility would be the square wave, which can be defined by

$$V_a = + V_0 \text{ for } \sin \omega t > 0$$
$$V_a = - V_0 \text{ for } \sin \omega t < 0.$$

Run the modified version of the program using various values for the parameters and discuss the results. Remember that it is only for a sinusoidally varying applied voltage that the applied voltage and the current have the same functional form (apart from a phase shift).

26. Nonlinear circuit elements.   Resistors, capacitors, and inductors are all linear circuit elements, in that the voltage across each element is proportional to the current through it. However, even for these circuit elements linearity is only an approximation which holds over some limited range of voltages and currents.

The diode is an example of a circuit element that is very nonlinear. For a typical semiconductor diode, the voltage V as a function of the current I is given by the formula

$$V = V_0 \ln (I_0 + I),$$

where the constants V_0 and I_0 depend on the particular diode. (This formula holds for both positive and negative values of the current I. However, it can only be used for negative currents which exceed $-I_0$.) It is still possible to define the resistance of such a circuit element in the usual way:

$$R = V/i = \frac{V_0 \ln (I_0 + I)}{I}, \tag{2.53}$$

although the resistance is clearly dependent on the current, rather than being a constant.

Modify the program so that it applies to the case of an RLC circuit in which R is a semiconductor diode rather than a linear resistor. This means that the function given in equation 2.53 must be used in place of a constant R which appears in several places in the program. When you select values for the constants V_0 and I_0, bear in mind that I_0 should be sufficiently large so that the argument of the logarithm in equation 2.53 is always positive. Run the modified version of the program using a sinusoidally varying applied voltage with several values for the parameters VO and W and discuss the shape of the resulting waveforms.

27. The damped harmonic oscillator.   The damped harmonic oscillator is the mechanical analog of the RLC series circuit. Let us assume that a mass m attached to a spring of force constant k is subject to the following three forces: a restoring force $-kx$, a viscous damping force $-\rho v$, and an applied force F_a. In this case, Newton's Second Law yields

$$-kx - \rho v + F_a = m \frac{dv}{dt}, \tag{2.54}$$

where

$$v = \frac{dx}{dt}. \tag{2.55}$$

Equations 2.54 and 2.55 are exactly the same form as equations 2.33 and 2.34, provided we make the substitutions

v   for I
x   for q
F_a for V_a
m   for L
ρ for R
k   for 1/C

In the present version of the RLC circuit program, the current I is calculated and plotted as a function of time. Modify the program to plot on the same graph, both the position $x (= q)$ and the velocity $v (= I)$ for a damped harmonic oscillator. Run the program using $F_a = 0$ and various values for the parameters m, ρ, and k. Make a further modification to use a damping force proportional to the square of the velocity: $F_d = \rho v^2$, which is the actual case for air resistance at moderate velocities. Another interesting case is $F_d = \rho$ (a damping force independent of velocity), which is approximately the case for sliding friction. Compare the position versus time graphs for the three types of damping.

28. Motion in one dimension.   The general problem of motion in one dimension can be solved using Newton's Second Law: $F = ma$, provided the force F is a known function of position, velocity, and time. To solve the problem numerically, we can use the improved Euler's method. The flow diagram shown in

Figure 2.26   Improved Euler's method for solving the problem of motion in one dimension

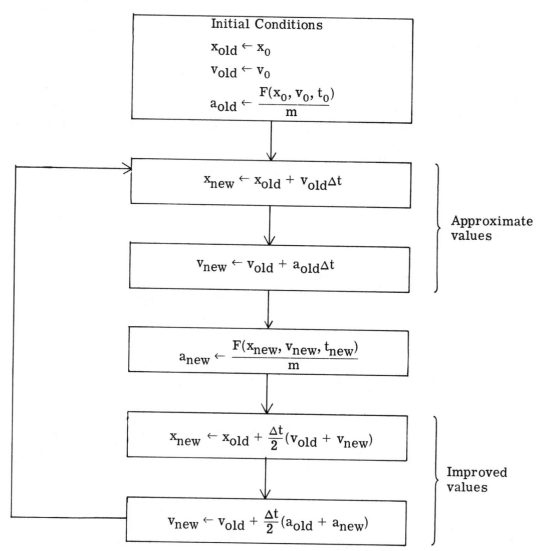

Figure 2.26 is very similar to that used in the RLC circuit problem in Fig. 2.25. We can, in fact, use the RLC circuit program to treat the problem of motion in one dimension, provided we make the change of variables suggested in problem 27. In the present case, since we are assuming that $F_a (= V_a)$ is the only force that acts, we must set $\rho (= R)$ and $k (= 1/C)$ equal to zero.

Modify the RLC circuit program, as suggested in problem 27, so that it plots both the position $x (= q)$ and velocity $v (= I)$ on the same graph. Run the program using various functional forms for the force F_a, and various values for the parameters. Some interesting possibilities are $F_a = F_0 \sin \omega t^2$ and $F_a = F_0 x^{-2}$. This last case applies to the one-dimensional motion of an object subject to the gravitational force of a single planet.

29. Motion in two and three dimensions.   To solve the general problem of motion in two or three dimensions, we can apply Newton's Second Law in vector form: $\mathbf{F} = m\mathbf{a}$. We note that each of the three component equations

$$F_x = ma_x \qquad F_y = ma_y \qquad F_z = ma_z$$

can be solved *independently,* when a numerical procedure is used. Thus, the problem of motion in three dimensions is no more complicated than three one-dimensional problems. The method of solving one-dimensional motion problems has already been discussed in problems 27 and 28.

Write a program to obtain the *trajectory,* i.e., the x, y, z coordinates of an object as a function of time, when the object is acted on by a force which is a known function of position, velocity, and time. Run the program using several functional forms for the force F. The simplest case is a constant F, which gives a parabolic trajectory if the initial velocity makes a nonzero angle with respect to the force. Other interesting cases include

$$\mathbf{F} = m\mathbf{g} + \rho v^2 \hat{\mathbf{v}},$$

which is the force acting on a projectile including air resistance, and

$$\mathbf{F} = \frac{K}{r^2} \hat{\mathbf{r}},$$

which is the inverse-square gravitational force acting on a satellite in the presence of a single planet ($\hat{\mathbf{v}}$ and $\hat{\mathbf{r}}$ are *unit vectors* along the **v** and r directions).

3

Waves in Classical and Quantum Physics

The Superposition Principle states that any number of waves can be combined into a resultant wave by adding the wave amplitudes at each point in space. This principle makes understanding wave phenomena in classical and quantum physics much simpler mathematically and physically. In this chapter, we investigate the superposition of waves, using the computer to obtain both numerical and graphical results. We also solve the Schrödinger equation for single and multiple square well potentials.

3.1   Superposition of Waves and Fourier's Theorem

We first consider the case of a wave whose amplitude can be represented by a periodic function $F(x, t)$ of one position variable x and the time t. We shall assume that the time variable has been fixed at some particular value, say t=0. Thus, since time is "stopped," we have a "snapshot" of the wave represented by the function $F(x) = F(x, 0)$, usually referred to as the *wave form*. Because $F(x)$ is a periodic function, for some value λ we must have $F(x + n\lambda) = F(x)$, for any integer n. By definition, λ is the wavelength of $F(x)$.

Of all possible periodic functions which might describe the shape of a wave form, the sine and cosine functions are most important. There are several reasons for the central role played by these particular functions. First, when any mechanical or electrical system undergoes small oscillations, they are usually of a sinusoidal nature. Examples of such mechanical systems include a mass on the end of a spring, a pendulum, a vibrating tuning fork, a cork bobbing in the water, and a high-rise building shaking in the wind. Examples of sinusoidally oscillating electrical systems include an LC circuit, a resonant microwave cavity, and an antenna. The waves emitted from sinusoidally oscillating systems also have a sinusoidal character, the wavelength λ being inversely proportional to the oscillation frequency of the system (and the waves).

A second reason for the importance of sine waves is that *any* periodic function $F(x)$ can be constructed by adding together some number of sine and cosine waves of different frequencies.* This result, known as Fourier's Theorem, can be expressed mathematically:

$$F(x) = \sum_{k=0}^{\infty} a_k \cos kx + \sum_{k=1}^{\infty} b_k \sin kx. \qquad (3.1)$$

---

* The requirement that $F(x)$ be periodic is *not* strictly necessary if we only wish to reproduce the function in some specific interval in x.

For simplicity, we have assumed that the function $F(x)$ is periodic, with wavelength 2π. For the case of an arbitrary wavelength λ, the argument of the sine and cosine functions in equation 3.1 would have to be changed to $2\pi(kx)\lambda$. The $k = 1$ term in the sine or cosine sums has wavelength and frequency equal to that of the function $F(x)$ itself, and the term is known as the *fundamental* or the *first harmonic*. The terms in the sine and cosine sums, sin kx and cos kx, for which $k = 2, 3, 4, \ldots$, are known as the second, third, fourth, $\ldots$, harmonics. The harmonics have progressively shorter wavelengths $\left(\dfrac{2\pi}{2}, \dfrac{2\pi}{3}, \dfrac{2\pi}{4}, \ldots\right)$, and therefore progressively higher frequencies, $(2, 3, 4, \ldots$ times the frequency of the fundamental). The coefficients a_k and b_k in equation 3.1 are the proportions of the kth sine and cosine waves needed to construct, or *Fourier synthesize,* the function $F(x)$. In most cases of interest, there are an infinite number of nonzero coefficients. However, in practice, a good approximation to the function $F(x)$ can often be obtained when the sums in equation 3.1 include only a finite number of terms. If we define a function $F_N(x)$ such that

$$F_N(x) = \sum_{k=0}^{N} a_k \cos kx + \sum_{k=1}^{N} b_k \sin kx, \tag{3.2}$$

then, by definition, we have $F(x) = \underset{N \to \infty}{\text{Lim}} F_N(x)$. We are interested in determining how the accuracy, or *goodness*, of the approximation of $F(x)$ by $F_N(x)$ depends on both the nature of the function $F(x)$ and the value of N.

If the coefficients $a_0, a_1, \ldots, a_N, b_1, b_2, \ldots, b_N$, are known, equation 3.2 can be used to evaluate $F_N(x)$ at a series of x-values, permitting a point by point comparison between the function $F(x)$ and the Fourier synthesized wave form $F_N(x)$. As shown in Appendix III, the coefficients a_k and b_k can be found using*

$$a_k = \frac{1}{\pi} \int_{-\pi}^{+\pi} F(x) \cos kx \, dx$$

and

$$b_k = \frac{1}{\pi} \int_{-\pi}^{+\pi} F(x) \sin kx \, dx. \tag{3.3}$$

Fourier Synthesis of Triangular, Square, and Spiked Wave Forms

The three wave forms displayed in Figure 3.1, which we shall refer to as a triangular wave, a square wave, and a spiked wave, can be defined in the interval $-\pi \leqslant x < \pi$, as follows:

triangular wave $\qquad F(x) = 1 - \dfrac{2}{\pi} |x| \qquad\qquad\qquad$ (3.4)

---

* These formulas hold for all positive integral k, $k = 1, 2, 3 \ldots$. For the coefficient a_0, the correct formula is

$$a_0 = \frac{1}{2\pi} \int_{-\pi}^{+\pi} F(x) \, dx.$$

This result can be obtained by integrating both sides of equation 3.1 between the limits $-\pi$ and $+\pi$. All terms in the sine and cosine sums integrate to zero except the $k = 0$ term which yields $2\pi a_0$.

square wave $\qquad$ $F(x) = +1$ for $0 \leqslant x < \pi$ $\qquad$ (3.5)

$\qquad\qquad\qquad\qquad$ -1 for $-\pi \leqslant x < 0$

spiked wave $\qquad$ $F(x) = \lim_{\epsilon \to 0} F_\epsilon(x)$, where $F_\epsilon(x) = \dfrac{1}{\epsilon}$ for $|x| < \dfrac{\epsilon}{2}$ $\qquad$ (3.6)

$\qquad\qquad\qquad\qquad$ 0 for $|x| > \dfrac{\epsilon}{2}$

Each of the three functions can be defined for all values of x, by noting that they are periodic and have a wavelength of 2π.

Figure 3.1   Three wave forms

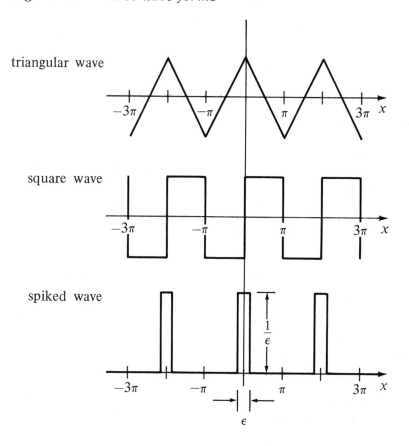

In Appendix IV we use equations 3.3 to find the coefficients a_k and b_k:

	a_k	b_k	
triangular wave	$\left(\dfrac{2}{\pi k}\right)^2 (1 - (-1)^k)$	0	(3.7)
square wave	0	$\left(\dfrac{2}{\pi k}\right)(1 - (-1)^k)$	(3.8)

| spiked wave* | $\dfrac{1}{N}$ | 0 | (3.9) |

The above formulas hold for all positive integral values of k, $k = 1, 2, 3, \ldots$.
The leading coefficient a_0 is taken to be zero in all three cases.** We see that
a_k is zero for the square wave and b_k is zero for the triangular and spiked
waves, for all values of k, from symmetry considerations. A symmetric function
(i.e., a function such that $F(-x) = F(x)$), such as a triangular or a spiked wave,
is expressible in terms of a sum involving only symmetric functions, such as
cosines. Similarly, an antisymmetric function (i.e., a function such that
$F(-x) = -F(x)$), such as the square wave, is expressible in terms of a sum involving only antisymmetric functions, such as sines. In the general case of a
function $F(x)$ which is neither symmetric nor antisymmetric, both the sine and
cosine sums are needed.

More on the Spiked Wave Form

For any wave form, the more waves we add in the Fourier synthesized wave form
$F_N(x)$, the more closely we approximate the wave form $F(x)$. In the case of the
spiked wave, this means that the width of the peak in the function $F_N(x)$ must become narrower as N increases, approaching zero as N approaches infinity. It
is instructive to see exactly how this comes about.

Figure 3.2   *$F_N(x)$ for a spiked wave, $N = 5$*

$$F_N(x) = \tfrac{1}{5} \ (cos \ x + \ cos \ 2x + \ cos \ 3x + \ cos \ 4x + \ cos \ 5x)$$

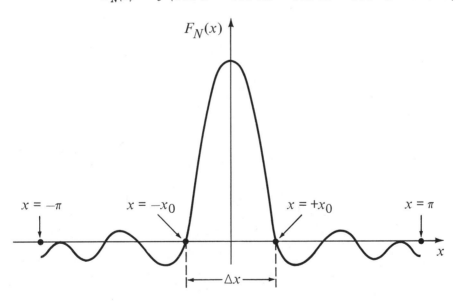

---

* The value obtained in Appendix IV for the spiked wave form shown in Figure
3.1 is $a_k = \dfrac{1}{\pi}$. We shall use the value $a_k = \dfrac{1}{N}$ instead, since all synthesized
spiked wave forms then have unit height which is convenient for plotting pur-
poses.

** This is equivalent to assuming that the wave form has zero net area between
$-\pi$ and $+\pi$. Clearly this is true for the square and triangular waves. It is
only true for the spiked wave if the wave form is moved down by an amount $\dfrac{1}{2\pi}$.

Let us define the *full width* Δx of the central peak in the function $F_N(x)$, as the distance between the first zeros at $x = \pm x_0$, on either side of the peak, i.e., $\Delta x = 2x_0$, where $F_N(\pm x_0) = 0$. (See Figure 3.2.) For the spiked wave, we have Fourier coefficients $a_1 = a_2 = a_3 = \ldots = a_N = \dfrac{1}{N}$ and $b_1 = b_2 = b_3 = \ldots = b_N = 0$, so that*

$$F_N(x) = \sum_{k=1}^{N} a_k \cos kx = \frac{1}{N}(\cos x + \cos 2x + \cos 3x + \cdots + \cos Nx). \quad (3.10)$$

This sum can be evaluated using N vectors, each of length $1/N$ arranged as shown in Figure 3.3. Each vector makes an angle x with the preceding one. The sum in equation 3.10 is equal to the sum of the horizontal components of the N vectors. We can find the smallest angle between the vectors which yields a zero resultant horizontal component by requiring the resultant vector to be vertical (see Figure 3.3b). In this case, the angle x between one vector and the next is given by

$$x = x_0 = \frac{\pi}{N+1}. \quad (3.11)$$

Thus, $x = x_0$ is the first zero in the function $F_N(x)$. The formula we have sought for the full-width of the peak is therefore

$$\Delta x = 2x_0 = \frac{2\pi}{(N+1)}. \quad (3.12)$$

Equation 3.12 is clearly consistent with the narrowing of the peak as N increases, the width approaching zero as N approaches infinity.

Figure 3.3   Sum of N vectors of length $\dfrac{1}{N}$ for N = 5

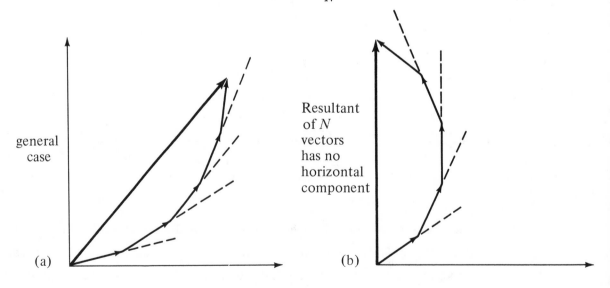

general case

(a)

Resultant of *N* vectors has no horizontal component

(b)

---

*   By omitting the $k = 0$ term in the sum, we are assuming the spiked wave has zero net area.

One interesting application of the spiked wave form is to the problem of a *wave packet,* which may be represented by a nonperiodic function which has a single spike and is zero everywhere else. This may be considered a limiting case of the spiked wave form, in which the wavelength or distance between spikes becomes infinite. Wave packets are of great utility in quantum mechanics, where this type of "localized" wave is used to represent a traveling particle. The width of the packet, Δx, is a measure of the uncertainty in the particle's position. The particle also has an uncertainty in its momentum Δp, which is proportional to the number of waves N, used in the Fourier synthesis of the wave packet. (This is because in quantum mechanics, a particle having a unique momentum p is represented by a sine (or cosine) wave of one specific frequency, so that a range of momenta Δp corresponds to a number of different frequencies.)

Putting the arguments of the preceding paragraph in mathematical form, we can use equation 3.12 to obtain one form of the Heisenberg Uncertainty Principle. For the kth harmonic, the momentum p_k can be found using the de Broglie relation

$$p_k = \frac{h}{\lambda_k},$$

where h is Planck's constant and

$$\lambda_k = \frac{2\pi}{k} \qquad \text{for } k = 1, 2, \ldots, N.$$

Thus

$$p_k = \frac{h}{2\pi} k,$$

giving

$$\Delta p = p_N - p_1 = (N - 1) \frac{h}{2\pi}. \tag{3.13}$$

If N is a large number, we may write $N - 1 \approx N + 1$, so that from equation 3.12 we obtain

$$N - 1 = \frac{2\pi}{\Delta x},$$

which, when substituted in equation 3.13, finally gives

$$\Delta p \Delta x = h,$$

which is one form of the Heisenberg Uncertainty Principle.*

Fourier Synthesis of Arbitrary Wave Forms

Fourier's Theorem states that *any* periodic function is expressible in terms of a sum of sine and cosine functions (equation 3.1). However, an algebraic ex-

---

*   This version of the Uncertainty Principle would normally be written as an inequality: $\Delta p \Delta x \geq h$, indicating that the uncertainties Δp and Δx may be larger than the values which satisfy $\Delta p \Delta x = h$, depending on the precision of the measuring apparatus.

pression for the coefficients a_k and b_k can only be found for certain functions (such as the square, triangular, and spiked wave forms). In many cases, the function $F(x)$ is such that an algebraic evaluation of the integrals in equation 3.3 is impossible. In other cases, even the function $F(x)$ itself may not be expressible in algebraic form. Should $F(x)$ represent some natural phenomenon, then the function can only be given in a graph or its equivalent, a table of numbers, giving the value of the function at a series of x-values (where x might represent time, rather than position). For example, to Fourier analyze human speech we must specify the sound intensity $F(x)$ at a series of times. In such cases, each of the coefficients $a_k, b_k, k = 1, 2, 3, \ldots N$, must be calculated using a numerical integration technique.

Goodness of the Approximation

The simplest way to judge the goodness of the approximation to a particular wave form, obtained using the Fourier expansion (equation 3.2) is to plot the function $F(x)$ and the Fourier synthesized function $F_N(x)$ on the same graph. An inspection of the two curves may be all that is necessary to determine if enough terms have been used to achieve the desired degree of closeness. To obtain a quantitative measure of the difference between the two functions $F(x)$ and $F_N(x)$, we calculate the *average deviation*, defined as

$$\Delta s_N = \frac{1}{M} \sum_{j=1}^{M} |F(x_j) - F_N(x_j)|, \tag{3.14}$$

where M is some number of points at which the functions $F(x)$ and $F_N(x)$ are sampled. If the function $F(x)$ is given only in terms of a table, M is the number of entries in the table. For a given function $F(x)$, by calculating Δs_N for a range of N values, we can see how the goodness of the approximation depends on the number of waves N.

The dependence of the goodness of the approximation on the number of waves can also be judged indirectly by seeing how a_k and b_k depend on k, since the faster a_k and b_k decrease with k, the faster Δs_N decreases with N. For the three functions depicted in Figure 3.1, we would expect the approximation to improve fastest for the triangular wave, for which $a_k \propto k^{-2}$, next for the square wave, for which $b_k \propto k^{-1}$, and slowest for the spiked wave, for which $a_k \propto k^0$. This result should not be surprising, in view of the fact that the three functions are increasingly "badly behaved" in this order. The triangular wave is everywhere continuous, but it has a discontinuous first derivative at the points $x = 0, \pm\pi, \pm 2\pi, \ldots$. The square wave has a discontinuity in the function itself, as well as the first derivative, at these points. Finally, the spiked wave has two discontinuities, at $x = \pm\epsilon$.

Computer Program for Fourier Synthesis

The computer program given in flow diagram form in Figures 3.5 and 3.6* and listed on pages 121-124, can compute and plot Fourier synthesized wave forms using specified numbers of waves. The program is capable of treating the three wave forms previously discussed (triangular, square, and spiked), as

---

* For problems such as this, involving complex flow diagrams, it is advisable not to cram everything onto a single page. Continuations from one page to another can be indicated using circled reference labels.

well as a fourth wave form shown in Figure 3.4 (a female profile), which has no algebraic form, but can be specified as a table of numbers obtained from the graph. The Fourier coefficients for this wave form can be found by evaluating the integrals in equations 3.3 using a numerical method such as Simpson's rule; the first 48 coefficients $a_1, \ldots, a_{48}; b_1, \ldots, b_{48}$ have been obtained in this way using another program and are specified in a DATA statement.

The program begins by reading from a data card numerical values for the parameters N and FLAG, where

N $\quad$ = number of waves in the Fourier synthesis
FLAG = 0, 1, 2, or 3 to indicate which of the four wave forms is desired: square, triangular, spiked, or female profile, respectively.

Any positive number may be used for N; however, for the case FLAG = 3, if N exceeds 48, only the first 48 waves will actually be included, since at present only the first 48 coefficients have been specified in the program. Note that if the value of N is 100 (for FLAG $\neq$ 3), then in addition to giving the usual output (the Fourier synthesized wave form), the program also generates a plot of the average deviations for N = 2, 4, 6, ..., N waves. This provides a quantitative measure of how the goodness of the approximation depends on the number of waves.

After reading the data card, the program stores the character ' ', (blank), in $Z_1, Z_2, \ldots, Z_{101}$. This array will later be used to produce the graph, line by line, using the scaling technique discussed on page 82. The program then sets the initial value of x = $-\pi$ and the increment $\Delta x = \dfrac{2\pi}{50}$. These values are chosen since we wish to calculate and plot the Fourier synthesized wave form at 51 equally spaced x-values between x = $-\pi$ and x = $+\pi$. The program next enters a loop over the 51 x-values. After initializing FN ($\equiv F_N(x)$) to zero, it next enters a loop over N, the number of waves. (This loop over the number of waves is contained within the loop over the 51 x-values.) For the first x-value, the

Figure 3.4   A fourth wave form

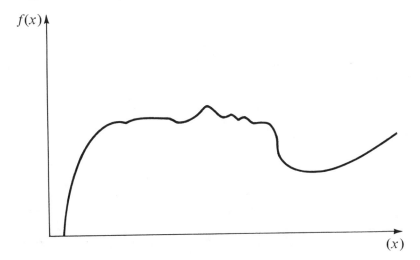

Figure 3.5   Flow diagram to Fourier synthesize wave forms—Part 1

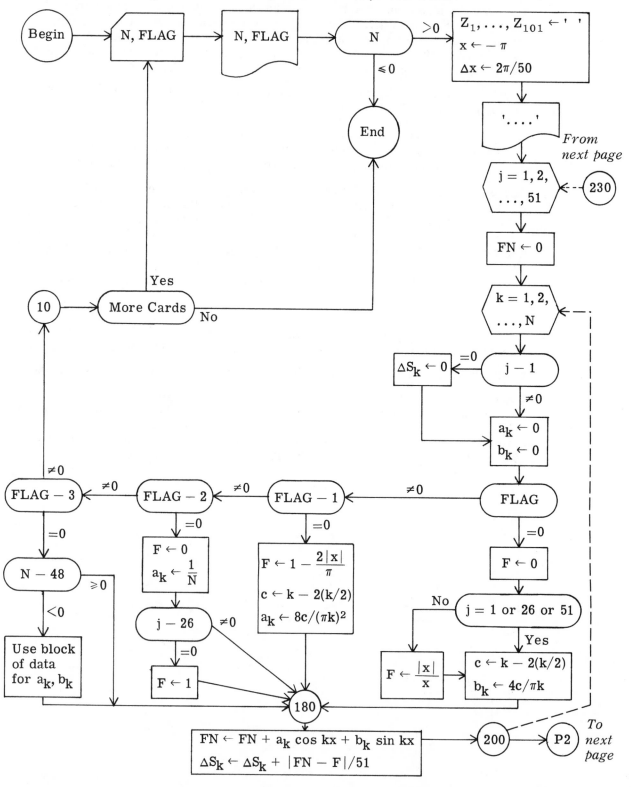

Figure 3.6   Flow diagram to Fourier synthesize wave forms—Part 2

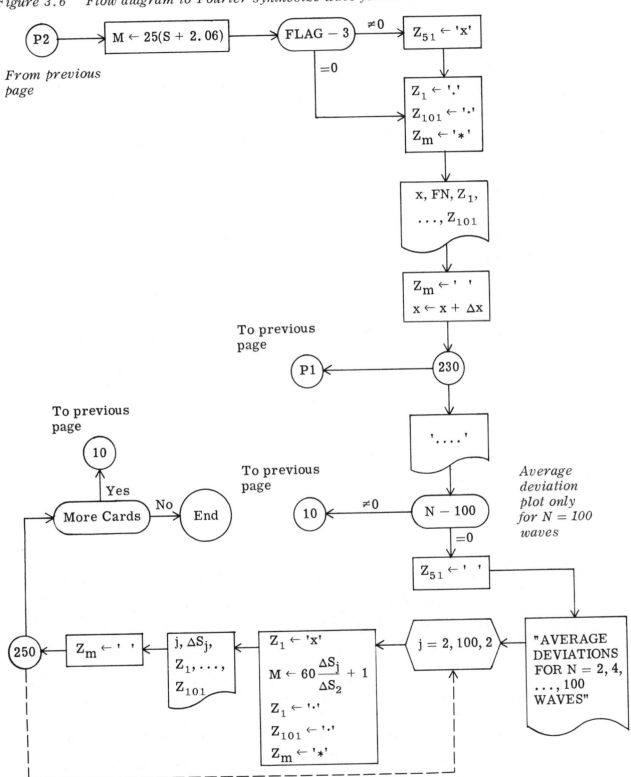

Figure 3.7

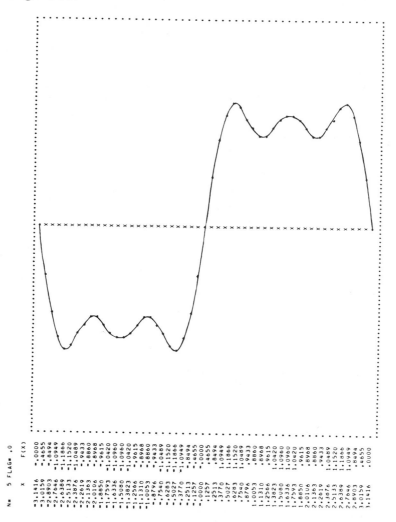

value of Δs_k, the average deviation for the Fourier synthesized wave form $F_k(x)$ is set to zero, initially. The program then tests the value of FLAG and calculates the coefficients a_k, b_k, and the function $F(x)$, according to the formulas that apply in each case for FLAG = 0, 1, 2, 3. If FLAG is not equal to one of these four values, the program reads a new data card (a "fail-safe" precaution). Having computed a_k, b_k, and $F(x)$, the program then adds the contributions of the kth wave (at the point x) to the sine and cosine sums, and if N = 100, it also adds the contribution to the average deviation Δs_k. Note that by computing the average deviation as each wave is added into the sum of sines and cosines, FN, we are able to obtain the deviations for k = 1, 2, 3, ..., N waves in a single run.

If we only wished to obtain numerical output, we could, at this point, print out the value of FN(=$F_N(x)$), and then proceed to the next x-value, until all 51 x-values between x = $-\pi$ and x = $+\pi$ have been done. However, since we wish to obtain a graph, as well as numbers, there is more to be done. (The flow chart

Figure 3.8

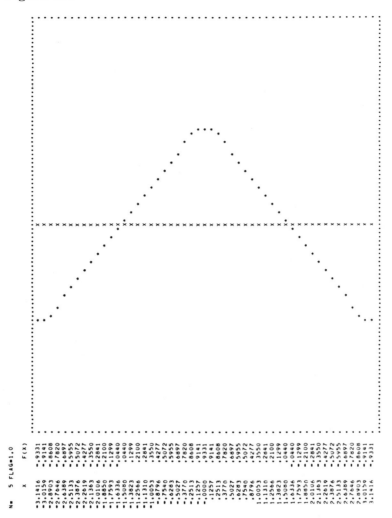

continues on Figure 3.6 at this point.) The program computes an integer M according to the linear scaling function:

$$M = 25 \ (FN + 2.06) \tag{3.15}$$

This choice of the scaling function restricts the values of M to the range $26 \leq M \leq 76$, corresponding to values of FN in the range $-1 \leq FN \leq +1$. If FLAG is not equal to 3 (the female profile), 'X' is stored in Z_{51} to mark the x-axis at the point corresponding to FN = 0. The edges of the graph are marked by storing a dot ('.') in Z_1 and Z_{101}. The program stores an asterisk ('*') in Z_M, so that the position of the asterisk, when the array $Z_1, Z_2, \ldots, Z_{101}$ is printed on a line, is proportional to the value of FN, according to the scaling function (equation 3.15). The program prints the numerical values of x and FN on the same line as $Z_1, Z_2, \ldots, Z_{101}$. It then resets Z_M back to ' ' (blank), increments x by Δx, and then repeats the whole process for the next value of x.

Figure 3.9

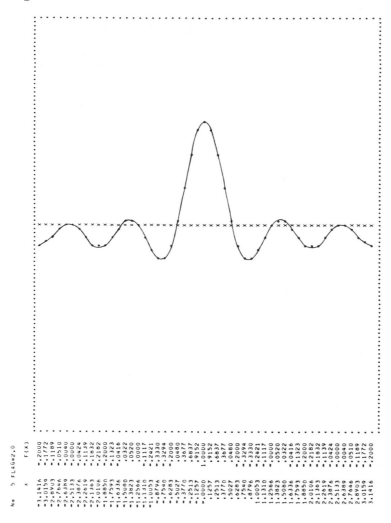

After FN has been calculated and plotted for 51 x-values, the graph, consisting of 51 printed lines, is complete.

If the value of N (the number of waves) is not equal to 100, there is nothing further to be done, and the program goes on to read the next data card, if any remain. If N equals 100, the program prints out a graph of the average deviations for each Fourier synthesized wave form $F_k(x)$, $k = 2, 4, 6, \ldots, 100$. (The only reason for doing every other wave is so that the graph can be made on a single page.) The graph is generated in a final loop. The scaling function used to make the graph of the deviations $\Delta s_2, \Delta s_4, \Delta s_6, \ldots, \Delta s_{100}$, is

$$M = 60 \frac{\Delta s_j}{\Delta s_2} + 1.$$

The form of this scaling function was chosen so that for Δs_2 (the largest

Figure 3.10

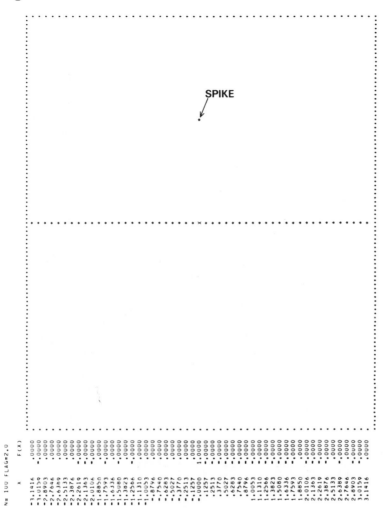

average deviation of the set), we have $M = 61$, and for the smallest possible deviation (zero), we have $M = 1$. After the program prints the numerical values and generates the graph of the average deviations, it reads the next data card if any remain.

Discussion of Sample Results

The sample results in Figures 3.7-3.12 were obtained using five data cards with the following values for the parameters:

	N	FLAG
1st card	5.0	0.0
2nd card	5.0	1.0
3rd card	5.0	2.0

Figure 3.11

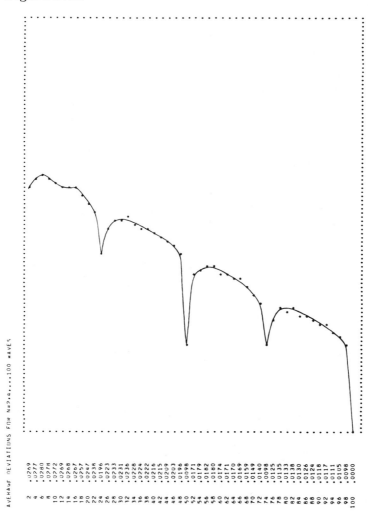

4th card	100.0	2.0
5th card	48.0	3.0

Recall that the values of FLAG on the first three cards specify the square, triangular, and spiked wave forms, respectively. The output for the fourth data card (Figure 3.10) is also for the spiked wave form; the large value of N causes the spike to be extremely narrow, as discussed on page 109 (it is so narrow that its width cannot be determined from this figure). Since the value of N is 100 for this case, the computer produces an additional graph (Figure 3.11) showing the average deviations for Fourier synthesized wave forms having $k = 2, 4, 6, \ldots,$ 100 waves. The output shown in Figure 3.12 is produced using the fifth data card on which N = 48 and FLAG = 3.

Problems 1-10 at the end of this chapter pertain to the superposition of waves and Fourier Theorem.

Figure 3.12

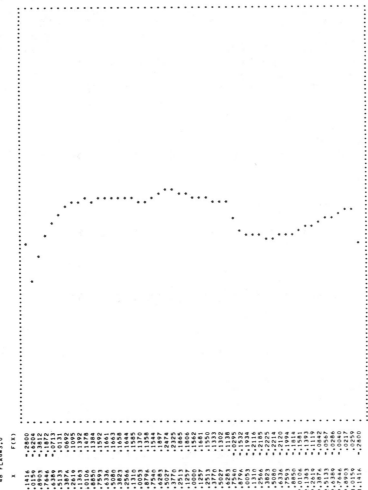

```
C           FOURIER SYNTHESIS OF VARIOUS WAVE FORMS
C
C●━━━━━━━━━━━━━━━━━━━━━━━━━━━━━━━━━━━━━━━━━━━━━━━━━━━━━━━━━━━━━━━━━
C
C          THIS PROGRAM DOES A FOURIER SYNTHESIS OF FOUR FUNCTIONS.  IT ADDS
C          SOME NUMBER OF SINE WAVES, EACH MULTIPLIED BY THE APPROPRIATE
C          COEFFICIENT, AND MAKES A PLOT OF THE RESULTANT WAVE FORM.
C          THE FOUR FUNCTIONS PRESENTLY PROGRAMMED ARE   SQUARE WAVE,
C          TRIANGLE,  SPIKE, AND PROFILE.
C
C          THE DATA STATEMENT DEFINES CHARACTERS NEEDED TO MAKE A PLOT.
C
      REAL N
      DIMENSION Z(101),DS(100),A(48),B(48)
      DATA BLANK,EX,ASTER,DOT/' ','X','*','.'/
      DATA A/   .1322,  .0138,  .0816,-.0360,  .0122,-.0379,-.0046,-.0147,
     1          .0089,  .0036,  .0084,  .0050,-.0027,-.0015,-.0055,  .0013,
     2         -.0042,  .0050,-.0017,  .0030,-.0041,  .0010,-.0047,-.0002,
     3         -.0019,  .0020,  .0001,  .0017,  .0002,-.0011,-.0016,-.0021,
```

```
    4           -.0001,-.0008,  .0023,  .0002,  .0015,-.0016,  .0003,-.0016,
    5            .0001,-.0002,  .0013,  .0004,  .0007,-.0003,-.0006,-.0009/
      DATA B/    -.1534,-.0468,  .0777,-.0440,  .0593,-.0544,  .0387,-.0457,
    1            .0348,-.0322,  .0381,-.0229,  .0306,-.0210,  .0205,-.0246,
    2            .0133,-.0202,  .0127,-.0130,  .0142,-.0103,  .0099,-.0121,
    3            .0066,-.0117,  .0075,-.0081,  .0088,-.0081,  .0079,-.0090,
    4            .0065,-.0093,  .0081,-.0081,  .0093,-.0074,  .0089,-.0092,
    5            .0076,-.0098,  .0074,-.0088,  .0091,-.0077,  .0084,-.0088/
C
C
C            READ A DATA CARD CONTAINING N AND FLAG
C            N IS THE NUMBER OF WAVES TO BE ADDED
C            FLAG=0,1,2,3
C            0 FOR SQUARE WAVE, 1 FOR TRIANGLE,  2 FOR SPIKE, 3 FOR PROFILE
C
   10   READ(2,1000)N,FLAG
        N1=N
        WRITE(3,1001)N1,FLAG
C
C
C            IF THE DATA CARD READ IN IS BLANK, N WILL BE ZERO AND WE QUIT.
C
        IF(N)260,260,20
   20   CONTINUE
C
C
C            WE PLOT 51 POINTS PER CYCLE, SO DX = 2*PI/50.
C
        PI=3.14159265
C
C
C            SET Z(1), Z(2), Z(3),...., Z(101) ALL EQUAL TO THE CHARACTER BLANK.
C
        DO 30 L=1,101
   30   Z(L)=BLANK
        X=-PI
        DX=2.0*PI/50.
        WRITE(3,1005)
C
C
C            LOOP OVER 51 POINTS MAKING ONE COMPLETE CYCLE.
C
        DO 230 J=1,51
C
C
C            THE SUM OF N WAVES IS INITIALLY SET TO ZERO.
C
        FN=0.
C
C
C            LOOP OVER THE NUMBER OF WAVES TO BE ADDED IN THE SUM.
C
        DO 200 K=1,N1
        IF(J-1)50,40,50
   40   DS(K)=0.
   50   XK=K
C
C
C            AK AND BK ARE THE FOURIER COEFFICIENTS.
C
        AK=0.
        BK=0.
        IF(FLAG)90,60,90
C
C
C                IF FLAG = 0, THEN
C                COMPUTE COEFFICIENTS FOR THE SQUARE WAVE.
C                F  IS THE THEORETICAL VALUE OF THE FUNCTION.
C
   60   F =0.
```

```
      AK=8.0*C/(PI*XK)**2
      GO TO 180
  110 IF(FLAG-2.0)140,120,140
C
C                IF FLAG = 2, THEN
C                COMPUTE COEFFICIENTS FOR SPIKE.
C
  120 F=0.
      AK=1.0/N
      IF(J-26)180,130,180
  130 F=1.0
      GO TO 180
C
C                COMPUTE COEFFICIENTS FOR PROFILE.
C
  140 IF(FLAG-3.0)170,150,170
C
C                CAN'T HANDLE MORE THAN 48 WAVES IN THIS CASE.
C
  150 IF(K-48)160,160,180
  160 AK=A(K)
      BK=B(K)
      GO TO 180
C
C                IF FLAG IS GREATER THAN 3, READ ANOTHER DATA CARD.
C
  170 GO TO 10
C
C                ADD THIS WAVE TO THE SUM.
C
  180 FN=FN+AK*COS(XK*X)+BK*SIN(XK*X)
C
C                COMPUTE AVERAGE DEVIATION FROM THEORETICAL FUNCTION.
C
      IF(N1-100)200,190,200
  190 DS(K)=DS(K)+ABS(F-FN)/51.
  200 CONTINUE
C
C                COMPUTE AN INTEGER M IN THE RANGE M=1,...,101 TO CORRESPOND TO
C                THE VALUE OF THE RESULTANT WAVE AMPLITUDE, STORED IN FN.
C                NOTE  THAT WHEN FN=0 WE HAVE M=51.
      IF((J-1)*(J-26)*(J-51))70,80,70
   70 F =ABS(X)/X
   80 CONTINUE
C
C                THE FACTOR C MAKES BK ZERO FOR EVEN VALUES OF K.
C
      C=K-2*(K/2)
      BK=4.0*C/(PI*XK)
      GO TO 180
   90 IF(FLAG-1.0)110,100,110
C
C                IF FLAG = 1, THEN
C                COMPUTE COEFFICIENTS FOR TRIANGLE.
C                THE FACTOR C MAKES AK ZERO FOR EVEN VALUES OF K.
C
  100 F =1.0-2.0*ABS(X)/PI
      C=K-2*(K/2)
      AK=8.0*C/(PI*XK)**2
      GO TO 180
  110 IF(FLAG-2.0)140,120,140
```

```
C
      M=25.*(FN+2.06)
C
C             STORE THE CHARACTER '*' IN Z(M).
C
C
C             MARK X AXIS, UNLESS FLAG=3.
C
      IF(FLAG-3.0)210,220,210
  210 Z(51)=EX
  220 CONTINUE
      Z(1)=DOT
      Z(101)=DOT
      Z(M)=ASTER
      WRITE(3,1002)X,FN,(Z(K),K=1,101)
      Z(M)=BLANK
C
C             INCREMENT X BY DX TO CALCULATE AND PLOT FN AT THE NEXT POINT.
C
  230 X=X+DX
      WRITE(3,1005)
C
C             GIVE DEVIATION PLOT ONLY FOR THE CASE N=100.
C
      IF(N1-100)10,240,10
  240 Z(51)=BLANK
      WRITE(3,1003)
      WRITE(3,1005)
      DO 250 J=2,100,2
      Z(1)=EX
      M= 60.*DS(J)/DS(2)+1.0
      Z(1)=DOT
      Z(101)=DOT
      Z(M)=ASTER
      WRITE(3,1004)J,DS(J),(Z(K),K=1,101)
  250 Z(M)=BLANK
      WRITE(3,1005)
      GO TO 10
 1000 FORMAT(2F10.5)
 1001 FORMAT(3H1N=,I4,6H FLAG=,F3.1//20H          X        F(X)    )
 1002 FORMAT(1X,F7.4,F9.4,2X,101A1)
 1003 FORMAT(42H1AVERAGE DEVIATIONS FOR N=2,4,...100 WAVES/)
 1004 FORMAT(1X,I3,F8.4,7X,101A1)
 1005 FORMAT(19X,101H.............................................................
     1.....................................................)
  260 CONTINUE
      CALL EXIT
      END
```

3.2 Waves in Two Dimensions

An example of a wave in two dimensions is the surface wave produced when a small object is dropped in calm water. The amplitude of the wave, which in this case is the vertical displacement of the water level from its equilibrium position, can be expressed as some function of two spatial coordinates x, y, and the time t. If the object bobs up and down in simple harmonic motion, it produces a sinusoidal wave which expands outward in all directions on the surface of the water (the xy-plane). As in the case of any *transverse* wave, the oscillatory motion of any point in the medium is perpendicular to the direction of propagation of the wave at that point. The amplitude of the wave at any point (x, y) and time t is given by

$$A(x, y, t) = \frac{S}{r} \sin \frac{2\pi}{\lambda} (r - ct), \tag{3.16}$$

where S indicates the strength of the source, r is the distance from the point source to the point (x, y), λ is the wavelength, and c is the velocity of the wave. At any particular instant of time, say t = 0, the wave is represented by the function

$$A(x, y) = \frac{S}{r} \sin \frac{2\pi}{\lambda} r. \tag{3.17}$$

This wave amplitude can be represented by the computer-generated plot shown in Figure 3.13 in which the darkness of the pattern at each point is an indi-

Figure 3.13

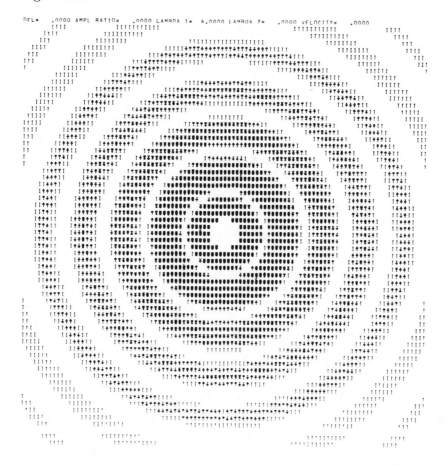

cation of the *intensity*, or square of the wave amplitude, at that point. (The procedure for doing this will be described.) Note that as required by equation 3.17, the wave amplitude is zero on concentric circles of radii $r = \dfrac{n\lambda}{2}, n = 0, 1, 2, \ldots,$ so that the distance between every other zero-amplitude circle is one wavelength. The alternating regions of positive and negative amplitude are indistinguishable in Figure 3.13 since it is the intensity, not the amplitude, that is represented. The figure does indicate both the sinusoidal oscillations as a function of r and the more gradual inverse square dependence of intensity on r, in accordance with the form of equation 3.17. Both Figure 3.13 and equation 3.17 apply only to the case of a point source which is at rest as far as motion in the xy-plane is concerned. In the next section we shall consider the case of a moving source.

Waves from a Moving Source: The Doppler Effect and Shock Waves

A *wave front* may be defined as the set of all points on the wave having the same phase. This means that the argument of the sine in equation 3.16 is the same for all points on a wave front. For a point source at rest, the wave fronts must therefore form a set of concentric circles spaced one wavelength apart, expanding outward at the wave speed c. If the source is not at rest, the wave fronts must still be circular, because the waves may be assumed to expand outward at the same velocity in all directions. However, the circles are not concentric, since adjacent circular wave fronts correspond to waves emitted from the source at different times, the center of each circle being the position of the source at the time that wave front was emitted from the source. For a source traveling at constant velocity, the center of each wave front is displaced by a constant amount from the one before. The computer-generated plot of Figure 3.14, which shows a snapshot of the wave pattern for a uniformly moving source, illustrates this.

The change in frequency which occurs when a wave source is in motion with respect to a stationary observer is known as the Doppler Effect. The frequency detected by the observer is higher (lower) than that for a stationary source, when the source moves toward (away from) the observer. The Doppler Effect also occurs when the observer, rather than the source, is moving, although this case will not be included in our discussion.

The cause of the Doppler Effect can be easily understood with the aid of Figure 3.14. Recall that the frequency, according to an observer, is the number of wave fronts reaching him per unit time. When a source approaches an observer (observer to the right of the source in Figure 3.14), the spacing between wave fronts is less than what it would be if the source were at rest. Similarly, if a source recedes from an observer, the spacing between wave fronts is greater than for a source at rest. The wave fronts reach the observer at the same velocity, whether the source is at rest or in motion. Therefore, the waves must have a higher (lower) frequency for a source which approaches (recedes from) an observer than for a source at rest.

In the simplest case, the source approaches an observer head-on with constant velocity. For this case, the observer detects a constant higher frequency, up to the instant the source passes him; after this instant, he detects a constant lower frequency. The frequency he detects as a function of the position of the source therefore looks like the "step function" curve (a) in Figure 3.15. This curve

Figure 3.14

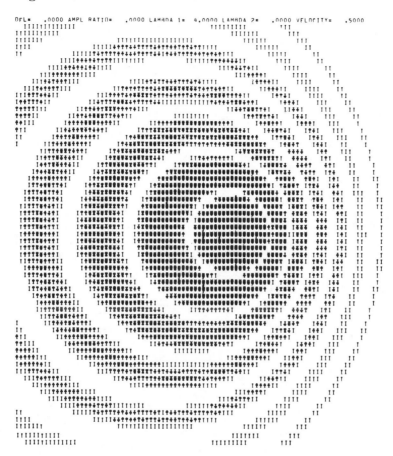

is consistent with the wave pattern in Figure 3.14. To the right of the source, the wave front spacing is about 1/2 the specified wavelength; to the left of the source, the spacing is about 3/2 times the wavelength. Since the observed frequency is inversely proportional to the wave front spacing, the two levels of the step function in Figure 3.15 correspond to frequencies which are respectively 2 and 2/3 times the frequency of the source at rest. Not surprisingly, these results based on a careful observation of Figure 3.14 agree with the formula* for the Doppler Effect for a moving source:

$$f = f_0/(1 \pm v/c),\tag{3.18}$$

where f is the frequency of the waves according to the observer, f_0 is the frequency the observer would detect if the source were at rest, and v/c is the ratio of the velocity of the source to the velocity of the waves. The plus sign

---

* The Doppler Effect formula is not used to generate the wave pattern. (For a description of the technique, see page 134.)

Figure 3.15   Frequency ratio f/f_0 versus position x

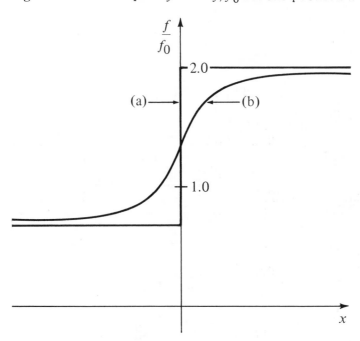

is for an approaching source, and the minus sign for a receding one. In the present numerical example, we have assumed $v/c = 1/2$.

A more difficult case to analyze arises when the source, still moving at constant velocity, does not approach the observer head-on. In this case, the frequency undergoes a more gradual change as the source passes the observer (see curve (b) in Figure 3.15). The detailed shape of this curve can be determined from the wave pattern in Figure 3.14. If we draw a horizontal line some distance above or below the horizontal through the center of the source we can measure the wave front spacing at a series of points along this line, and from this the observed frequency at each point. Here we assume that the observer travels along the horizontal line through the wave pattern. This is equivalent to our original assumption of a stationary observer and a moving source, in which case the wave pattern depicted would move horizontally past the observer.

A shock wave is produced when the source of waves travels at a speed v, which exceeds the speed of the waves c. As shown in the computer-generated snapshot in Figure 3.16, the wave amplitude is zero outside a certain angular region (a cone in the three-dimensional case). This effect could manifest itself as the wake of a boat in the case of water waves, a sonic boom in the case of sound waves, or Cerenkov radiation in the case of light waves. The shock wave arises due to the superposition of waves emitted from the moving source at different times. Figure 3.17 shows a series of wave fronts emitted at five instants of time when the source is located at points A, B, C, D, and E. Constructive addition of the five wave fronts (as well as all others) occurs along a direction θ shown as

Figure 3.16

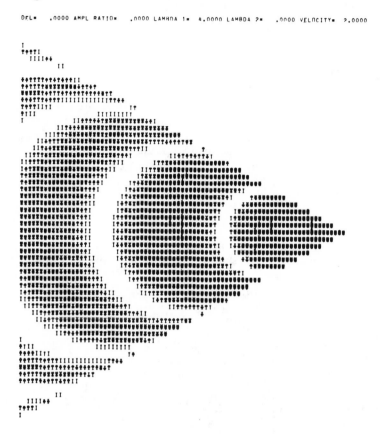

Figure 3.17   Wave fronts emitted from a moving source

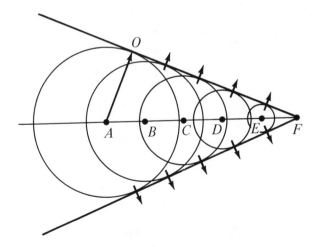

the angle AFO.* The line OF forms one edge of the shock wave. We see that the ratio $AO/AF = c/v$, from which we obtain

$$\sin \theta = c/v. \tag{3.19}$$

Superposition of Waves in Two Dimensions

Sources are *coherent* if their relative phase does not vary in time. Two coherent point sources of waves will produce an interference pattern, such as that in Figure 3.18, as a consequence of the Superposition Principle according to which the resultant wave amplitude at any point in the xy-plane is equal to the sum of the individual wave amplitudes, $A_1 + A_2$, at that point. An interference pattern arises, since the relative phase between A_1 and A_2 varies from point to

Figure 3.18

```
DFL= 4.0000 AMPL RATIO= 1.0000 LAMBDA 1= 1.5000 LAMBDA 2= 1.5000 VELOCITY=  .0000
```

---

* Shock waves for sound or light are three-dimensional, so that θ is then actually the angle of a cone, the axis of which is along the direction of motion of the source (the "Mach cone").

point, due to the changing distances to each of the sources. Given two coherent point sources, separated by a distance d, which emit waves that are initially in phase, we may write for the amplitudes at a point (x, y), at time t = 0:*

$$A_1(x, y) = \frac{S_1}{r_1} \sin \frac{2\pi}{\lambda_1} r_1$$

$$A_2(x, y) = \frac{S_2}{r_2} \sin \frac{2\pi}{\lambda_2} r_2,$$

(3.20)

where S_1 and S_2 are the source strengths, λ_1 and λ_2 are the wavelengths, and r_1 and r_2 are the distances from each source to the point (x, y). The technique used to generate an interference pattern, such as that in Figure 3.18, is to divide a region of the xy-plane using a rectangular grid. At each grid point, after calculating the individual wave amplitudes using equation 3.20, we then find the resultant amplitude $A = A_1 + A_2$, and finally, the resultant intensity, $I = A^2$.

In the inteference pattern in Figure 3.18, the sources are assumed to have equal strengths and equal wavelengths. In the usual treatment of this problem, the condition for destructive interference is that the waves from one source travel a distance which is an odd integral number of half-wavelengths more than the waves from the other source, i.e., $r_1 - r_2 = n\frac{\lambda}{2}$, n = 1, 3, 5, ... However, this unrealistically ignores the inverse square dependence of intensity on distance for each source. The computer-generated interference patterns can be made, as indicated in the preceding section, without using any such simplifying assumption. Also, notice in Figure 3.18 the series of maxima and minima along the line joining the sources; this phenomenon would not arise in the usual r-independent treatment. It would be expected, however, that at distances from each source that are large compared to the separation between sources d, there would be little difference between the r-dependent and r-independent treatment. In particular, we would expect that at distances from the sources that are large compared to their separation, the directions θ with respect to the line joining the sources, for which maxima and minima in the intensity are found, are given by the conditions:

maxima: $n\lambda = d \cos \theta$,   minima: $(n + \frac{1}{2})\lambda = d \cos \theta$,

where n is any integer.

For two sources of unequal strengths (see, for example, Figure 3.19), the interference maxima and minima are more pronounced on the side of the weaker source, where, due to the inverse square dependence of intensity on distance, the two wave amplitudes tend to have more equal magnitudes. The directions along which minima and maxima in the intensity occur at large distances are the same as for the case of equal source strengths. Thus, the points of maximum and minimum intensity lie on straight lines, except near the sources. As we might expect, this is not the case for two sources having different wavelengths (see Figure 3.20).

---

*   The case of two incoherent sources, i.e., sources having a constantly changing relative phase, is treated in problem 15.

Figure 3.19

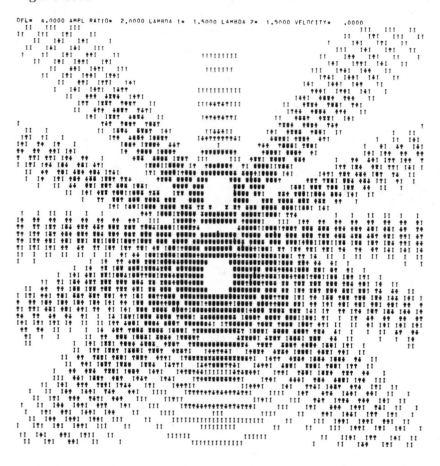

Two-Dimensional Fourier Analysis*

The method of Fourier analysis and synthesis of wave forms can easily be extended to treat two-dimensional waves. For an arbitrary function of x and y which is periodic in both variables, we may write

$$F(x, y) = \sum_{k=0}^{\infty} \sum_{l=0}^{\infty} a_{kl} \cos (kx + ly) + \sum_{k=1}^{\infty} \sum_{l=1}^{\infty} b_{k1} \sin (kx + ly) \qquad (3.21)$$

The functions cos (kx + ly) and sin (kx + ly) represent *plane waves,* i.e., waves having wave fronts that are planes (straight lines in two dimensions). Their wavelength is $(k^2 + l^2)^{-1/2}$, and they have wave fronts making an angle $\theta = \tan^{-1}(k/l)$ with the x-axis. If we interpret the function F(x, y) as the intensity at

---

*   This section may be skipped on a first reading. The present version of the wave pattern program does not treat the problem of two-dimensional Fourier synthesis. (However, see problem 19.)

Figure 3.20

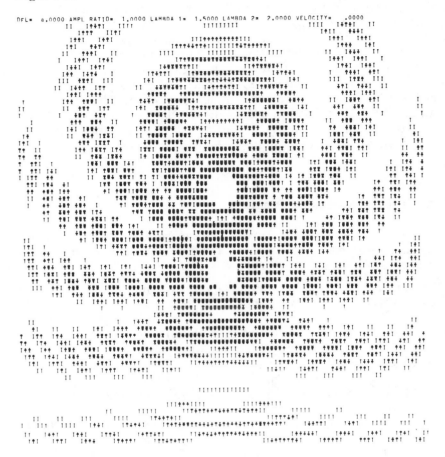

a point (x, y), we can then use the wave pattern program (with some modification —see problem 19) to generate patterns corresponding to particular choices of the Fourier coefficients $a_{kl}, b_{kl}, k = 1, 2, 3, \ldots, 1 = 1, 2, 3, \ldots$.

A recently developed practical application of the two-dimensional Fourier analysis technique involves the analysis and reconstruction of photographs or other pictorial data in order to enhance or suppress particular features. While the intensity variations in a picture are generally of a nonperiodic nature, we can use equation 3.21 to do a Fourier synthesis by choosing the lowest frequency component to have a wavelength equal to the picture size, so that the function $F(x, y)$ repeats itself only outside the boundaries of the picture. There are usually four steps in the Fourier analysis and reconstruction of photographs by computer: (1) Make measurements of the darkness of a picture at a finite grid of points to obtain the function $F(x, y)$ in numerical form from the picture. (2) Calculate the "two-dimensional" Fourier coefficients a_{kl} and b_{kl} from the function $F(x, y)$ using equations analogous to equations 3.3, but involving two-dimensional integrals. (3) Adjust the Fourier coefficients so as to either enhance desired features or suppress unwanted features in the picture. (This step is quite similar to adding some treble or bass to give a more desirable frequency response in a hi fi system). (4) Reconstruct the picture point-by-point using the adjusted coefficients in equation 3.21 with some finite number of terms (Fourier components). Two examples of this technique are shown in Figures 3.21 and 3.22. Figure 3.21a is a photograph of the lunar surface from an unmanned Surveyor mission. Figure 3.21b is the reconstructed picture in which glare has been largely eliminated through the partial removal of the low-frequency Fourier components. Figure 3.22a is a portion of a picture of the lunar surface from an unmanned Ranger mission. The original picture transmitted back to earth has a noticeable amount of sinusoidal interference (noise). This noise is largely absent in the reconstructed picture (Figure 3.22b) through suppression of anomalously large Fourier components.

Wave Pattern Computer Program

The wave pattern computer program given in flow diagram form in Figures 3.24 and 3.25 and listed on pages 143-145, can be used to generate all the wave patterns in Figures 3.13-3.20.

The program first reads numerical values from a data card for the parameters DEL, RATIO, L1, L2, and V, where

DEL $= \Delta$ (separation between sources)
RATIO $= S_2/S_1$ (ratio of source strengths)
L1 $= \lambda_1$ (wavelength for source 1)
L2 $= \lambda_2$ (wavelength for source 2)
V $= v/c$ (ratio of the source and wave velocities).

To generate the single source patterns (Figures 3.13, 3.14, and 3.16), RATIO must be set to zero, and to generate the interference patterns for two stationary sources (Figures 3.18, 3.19, and 3.20), V must be set to zero.

After the data card has been read, the program proceeds to calculate the resultant intensity (RINT) at all points within a rectangular grid of points inside the square region defined by $-10 \leqslant x \leqslant +10, -10 \leqslant y \leqslant +10$. The points in the grid are identified by a pair of indices (j, k), where $j = 1, 2, \ldots, 61$ indicates the row number, and $k = 1, 2, \ldots, 101$ indicates the column number. At each grid point starting with $(j, k) = (1, 1)$, the program first computes the coordinates (x, y) in terms of the indices (j, k), using

$$x = \tfrac{1}{5}(k - 51)$$
$$y = \tfrac{1}{3}(j - 31).$$

(3.22)

If the two sources are at rest $(V = 0)$, they are assumed to be symmetrically located, a distance Δ apart at the points $\left(0, -\dfrac{\Delta}{2}\right), \left(0, +\dfrac{\Delta}{2}\right)$ (see Figure 3.23). In this case, we can use the distance formula to obtain r_1 and r_2, the distances from each source to the point (x, y):

$$r_1 = \left(x^2 + \left(y - \frac{\Delta}{2}\right)^2\right)^{1/2} \quad (3.23) \qquad r_2 = \left(x^2 + \left(y + \frac{\Delta}{2}\right)^2\right)^{1/2} \quad (3.24)$$

Figure 3.21
(a) Surveyor photograph of lunar surface (Photos courtesy of JPL)

Figure 3.21   (b) Reconstructed picture

Figure 3.22 (a) Ranger photograph of lunar surface

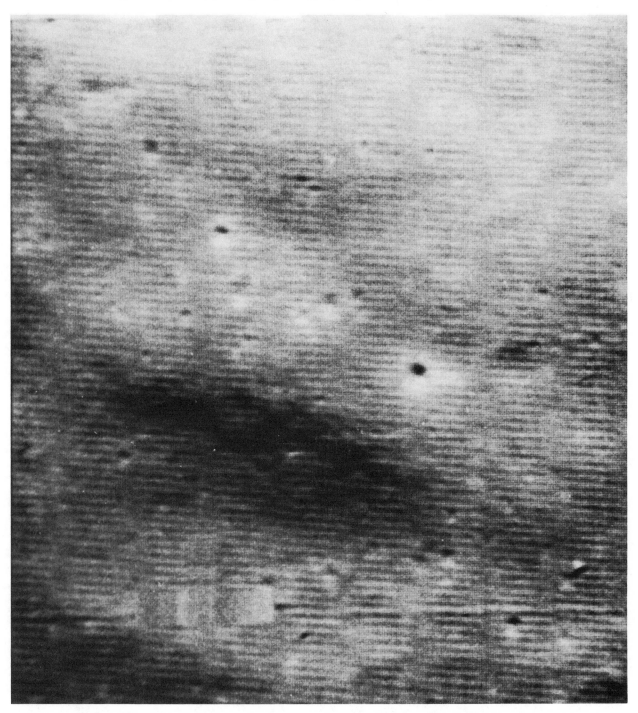

Figure 3.22   (b) Reconstructed picture

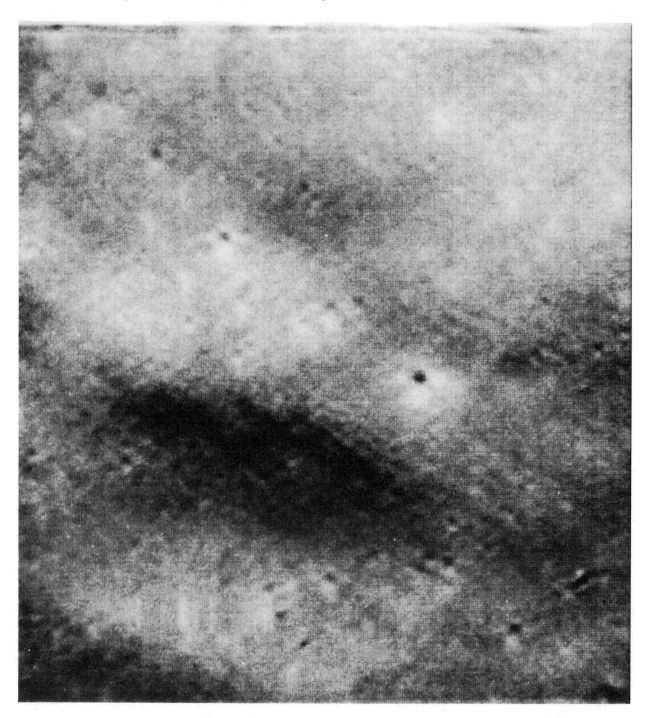

If source 1 is in motion, r_2 can still be found using equation 3.24, but the procedure for finding r_1 is more complicated. We assume that the source moves along the x-axis at constant speed and that at time t = 0 (the instant at which a wave front reaches the point (x, y)), the source is located at a particular point (D, 0). We need to know the location of the source at the time when the wave front reaching (x, y) at t = 0 *was emitted from the source* (see Figure 3.23). The technique for finding the location of the source $(x_0, 0)$ at the emission time is described in Appendix V. Once the value of x_0 has been determined, r_1 can then be obtained using the distance formula

$$r_1 = ((x - x_0)^2 + y^2)^{1/2}. \tag{3.25}$$

There is, however, one important exception. It may happen that no solution for x_0 exists. This occurs when v exceeds c, and the point (x, y) is outside the Mach cone. In this event, the amplitude of the wave at the point (x, y) is zero. In order to treat this case in the program, using the same formula for the amplitude (equation 3.20), the program sets $r_1 = \lambda_1$, which achieves the desired result, because then $\sin (2\pi r_1/\lambda_1) = 0$.

Once the program obtains r_1 and r_2 using either equations 3.23 and 3.24 (if both sources are at rest), or equations 3.24 and 3.25 (if source 1 is in motion), it adds .000001 to r_1 and r_2, so that when it computes the wave amplitudes A_1

Figure 3.23   *Locations of two point sources assumed in program*

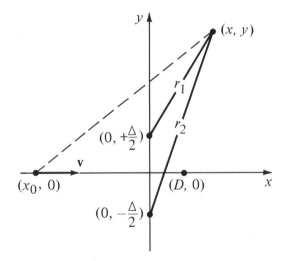

r_2 = distance from source 2 to point (x,y)

r_1 = distance from source 1 to point (x,y) if v = 0.

The dotted line shows r_1 if v ≠ 0.

(D, 0) is the location of source 1 at time t = 0.

Figure 3.24   Flow diagram to generate wave patterns—Part 1

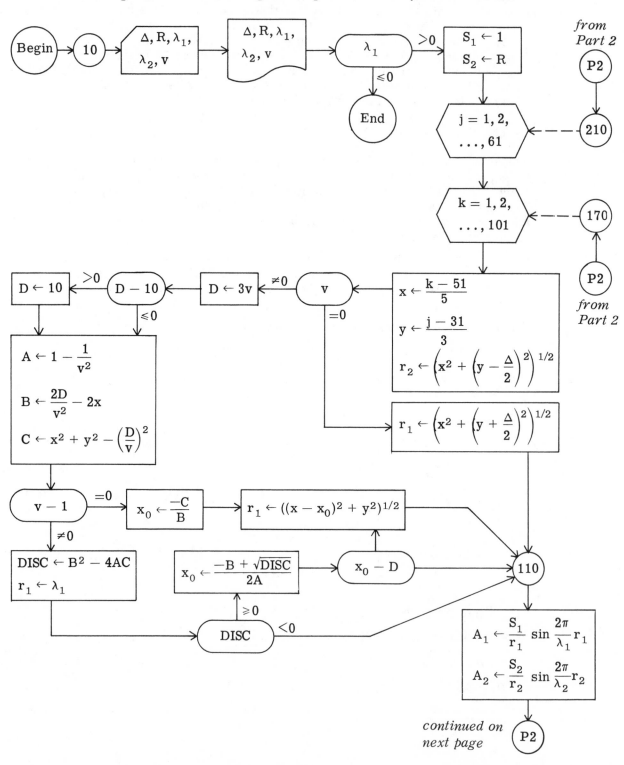

continued on
next page

Figure 3.25   Flow diagram to generate wave patterns—Part 2

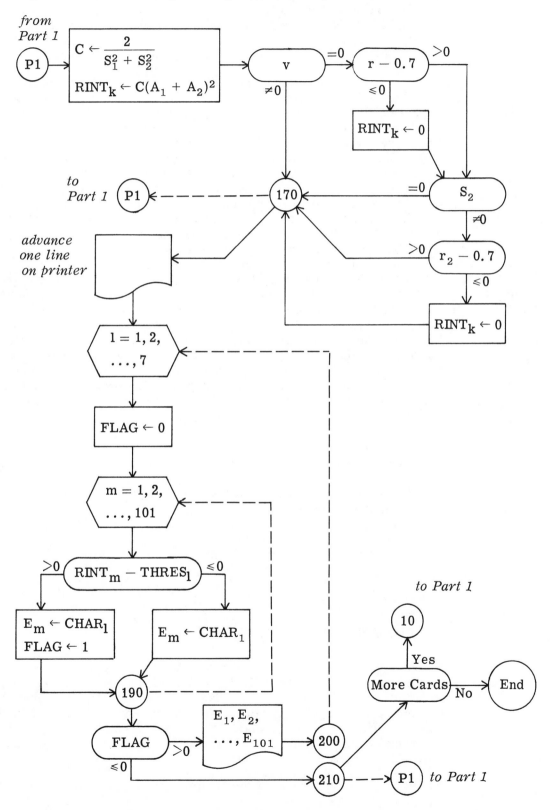

and A_2 according to equation 3.20, it will never divide by zero. It finally computes the **resultant intensity** RINT $= A^2 = (A_1 + A_2)^2$. For reasons which will be clear later, the program sets RINT $= 0$ if the point (x, y) is very close to either source and V is not zero. (This produces small blanked-out circular regions around each source, which serve to locate each source.)

After finding the resultant intensity at one point, the program proceeds to the next point within that row. Once all the points within a row have been considered, the array RINT(K), K $= 1, 2, \ldots, 101$, contains the resultant intensities. The program then loops through the array RINT seven times. Each time, it compares every element of the array RINT to a "threshold intensity," THRES(L), to determine if the intensity is above some threshold level. If the intensity RINT(K) is above threshold, a character CHAR(L) is stored in the Kth position of the array E, and if it is below threshold, the character blank is stored. Once all 101 positions within a row have been assigned a CHAR(L), the array (E(K), K $=$ 1, 101) is printed. The array appears as a row of characters, with blanks occurring at positions where the intensity is below some threshold value THRES(L). Since the program loops through the array RINT seven times, the procedure just described is carried out seven times, using a progressively higher threshold each time. The seven lines are printed *directly on top of one another (overprinted)*. (This is accomplished by using a plus sign as a control character in the FORMAT statement.) Thus, the highest intensity points will be represented by seven different characters printed on top of one another, and lower intensity points will be represented by fewer overprinted characters (possibly none). At each position in the pattern, the number of overprinted characters, and therefore the darkness of the pattern, is determined by the value of the resultant intensity at that point. The program produces the pattern, row by row, until all 61 rows have been printed, and the pattern is then complete.

Comments on Data Cards and Sample Results

The data cards used to generate Figures 3.13-3.19 contained the following values for the parameters:

	DEL	RATIO	L1	L2	V
Fig. 3.13	0.	0.	4.0	0.	0.
Fig. 3.14	0.	0.	4.0	0.	0.5
Fig. 3.16	0.	0.	4.0	0.	2.0
Fig. 3.18	4.0	1.0	1.5	1.5	0.
Fig. 3.19	4.0	2.0	1.5	1.5	0.
Fig. 3.20	4.0	1.0	1.5	2.0	0.

It should be noted that it is time-consuming for the computer to generate each plot (about ten minutes per plot for an IBM 1130). For this reason, it may be impractical for each student in a class to generate a large number of such plots. One possibility would be for the instructor to generate a large number of plots and distribute copies to the class. Each student might then try to determine the numerical values of each of the parameters DEL, RATIO, L1, L2, and V, based on a careful examination of each plot.

Problems 11-19 at the end of this chapter pertain to waves in two dimensions.

```
C             WAVE PATTERNS
C
C-------------------------------------------------------------------------
C
C             THIS PROGRAM DISPLAYS INTERFERENCE PATTERNS WHICH ARE CALCULATED
C             FOR TWO POINT SOURCES OF WAVES.   THE WAVE AMPLITUDE FROM EACH
C             SOURCE IS ASSUMED TO BE INDEPENDENT OF ANGLE, AND IT VARIES AS THE
C             INVERSE FIRST POWER OF THE DISTANCE FROM THE SOURCE.
C
      REAL L1,L2
      DIMENSION E(101),CHAR(7),RINT(101),THRES(7)
C
C             THIS DATA STATEMENT DEFINES CHARACTERS NEEDED TO MAKE A PLOT.
C
      DATA CHAR/' ','I','-','X','O','.','M'/
C
C             THIS DATA CARD DEFINES A SET OF INTENSITY THRESHOLDS.
C
      DATA THRES/.005,.01,.02,.04,.08,.16,.32/
      PI=3.14159265
C
C             A DATA CARD IS READ CONTAINING DEL, RATIO, L1, L2, AND V.
C
C             DEL  = SEPARATION BETWEEN SOURCES
C             RATIO= RATIO OF TWO SOURCE STRENGTHS
C             L1,L2= WAVELENGTHS
C             V    = VELOCITY OF SOURCE 1.(VELOCITY OF WAVES IS ASSUMED TO BE 1)
C
   10 READ(2,1000)DEL,RATIO,L1,L2,V
      WRITE(3,1001)DEL,RATIO,L1,L2,V
C
C             IF THE DATA CARD READ IN IS BLANK, THEN L1 IS ZERO, AND WE QUIT.
C
      IF(L1)220,220,20
   20 CONTINUE
C
C             THE FIRST  OF THE TWO SOURCE STRENGTHS WILL BE ARBITRARILY
C             ASSUMED TO BE 1.0, THE SECOND  HAS THE VALUE OF RATIO.
C
      S1=1.0
      S2=RATIO
C
C             THIS LOOP GOES OVER 61 ROWS.
C
      DO 210 J=1,61
C
C             THIS LOOP GOES OVER 101  COLUMNS WITHIN A ROW.
C
      DO 170 K=1,101
C
C             THE X AND Y COORDINATES ARE CALCULATED FROM THE ROW AND COLUMN
C             INDICES J AND K.
C
      AJ=J
      AK=K
      X=(AK-51.)/5.
      Y=(AJ-31.)/3.
C
```

```
C                    R1 AND R2 ARE THE DISTANCES FROM EACH SOURCE TO THE POINT (X,Y).
C
      R2=SQRT(X**2+(Y-DEL/2.)**2)
      IF(V)40,30,40
   30 R1=SQRT(X**2+(Y+DEL/2.)**2)
      GO TO 110
C
C                    THE STATEMENTS BETWEEN THE DOTTED LINES ONLY APPLY IF THE
C                    VELOCITY OF SOURCE 1 IS NONZERO.
C
C----------------------------------------------------------------------
   40 D=3.0*V
      IF(D-10.0)60,60,50
   50 D=10.0
   60 A=1.0-1.0/V**2
      B=2.0*D/V**2-2.0*X
      C=X**2+Y**2-(D/V)**2
      IF(V-1.0)80,70,80
   70 X0=-C/B
      GO TO 100
   80 CONTINUE
      DISC=B**2-4.0*A*C
      R1=L1
      IF(DISC)110,90,90
   90 X0=(-B+SQRT(DISC))/(2.0*A)
      IF(X0-D)100,100,110
  100 R1=SQRT((X-X0)**2+Y**2)
  110 CONTINUE
C----------------------------------------------------------------------
C
C                    ADD .000001 TO R1 AND R2 TO AVOID DIVIDING BY ZERO.
C
      R1=R1+.000001
      R2=R2+.000001
C
C                    A1 AND A2 ARE THE AMPLITUDES OF EACH WAVE AT THE POINT (X,Y).
C
      A1=S1*SIN(2.*PI*R1/L1)/R1
      A2=S2*SIN(2.*PI*R2/L2)/R2
C
C                    RINT IS THE RESULTANT INTENSITY OF THE TWO WAVES COMBINED,
C                    CONS IS AN ARBITRARY NORMALIZATION CONSTANT.
C
      CONS=2.0/(S1**2+S2**2)
      RINT(K)=CONS*(A1+A2)**2
C
C                    IF THE POINT (X,Y) IS CLOSER THAN 0.7 UNITS TO EITHER SOURCE,
C                    SET THE INTENSITY DIGIT TO 0, TO MAKE A CIRCULAR BLANK REGION
C                    AROUND THE LOCATION OF EACH SOURCE.
C
      IF(V)170,120,170
  120 IF(R1-0.7)130,130,140
  130 RINT(K)=0.
  140 IF(S2)150,170,150
  150 IF(R2-0.7)160,160,170
  160 RINT(K)=0.
  170 CONTINUE
C
```

```
C           ADVANCE TO THE NEXT LINE ON THE PRINTER WITHOUT ACTUALLY WRITING
C           ANYTHING.
C
      WRITE(3,1003)
C
C           THIS LOOP MEANS THAT EACH LINE OF OUTPUT IS OVERPRINTED 7 TIMES.
C
      DO 200 L=1,7
C
C           THIS LOOP GOES OVER THE 101  COLUMNS WITHIN A GIVEN ROW.
C
      FLAG=0.
      DO 190 M=1,101
C
C           IF INTENSITY EXCEEDS THRESHOLD VALUE THRES(L), REPRESENT IT BY
C           THE CHARACTER CHAR(L), OTHERWISE USE CHAR(1) WHICH IS A BLANK.
C
      IF(RINT(M)-THRES(L))185,180,180
  180 E(M)=CHAR(L)
      FLAG=1.0
      GO TO 190
  185 E(M)=CHAR(1)
  190 CONTINUE
      IF(FLAG)210,210,200
  200 WRITE(3,1002)(E(K),K=1,101)
  210 CONTINUE
C
C           READ NEXT DATA CARD TO MAKE ANOTHER INTENSITY PATTERN.
C
      GO TO 10
 1000 FORMAT(5F10.5)
C
C           TO OVERPRINT A LINE, THE FORMAT STATEMENT SHOULD HAVE A '+' AS
C           A CONTROL CHARACTER.
C
 1001 FORMAT(5H1DEL=, F8.4,12H AMPL RATIO=, F8.4,10H LAMBDA 1=, F8.4,
     1 10H LAMBDA 2=, F8.4,10H VELOCITY=, F8.4)
 1002 FORMAT(1H+,101A1)
 1003 FORMAT(1H )
  220 CONTINUE
      CALL EXIT
      END
```

3.3 Solution of the Schrödinger Equation

According to the principles of quantum mechanics, the probability of finding a particle at some point in space and time is proportional to the square of the absolute value of its wave function $\psi(x, y, z, t)$ at that point. To find the wave function for a nonrelativistic particle of energy E subject to a potential V, we must solve the Schrödinger equation. For the one-dimensional case ($V = V(x)$, $\psi = \psi(x, t)$), the time independent Schrödinger equation can be written

$$\frac{\hbar^2}{2m} \frac{d^2\psi}{dx^2} = (V(x) - E)\psi. \tag{3.26}$$

We shall for simplicity use a system of units in which Planck's constant $\hbar = 1$ and the mass of the particle $m = \frac{1}{2}$. Equation 3.26 can therefore be written

$$\frac{d^2\psi}{dx^2} = (V(x) - E)\psi. \tag{3.27}$$

The Square Well Potential

A square well potential can be defined by the conditions

$V(x) = V_0$ inside the well $(0 < x < w)$

$V(x) = 0$ outside the well,

where V_0 (a *negative* number) and w are the depth and width of the well, respectively. (See Figure 3.26.)

Figure 3.26   The square well potential

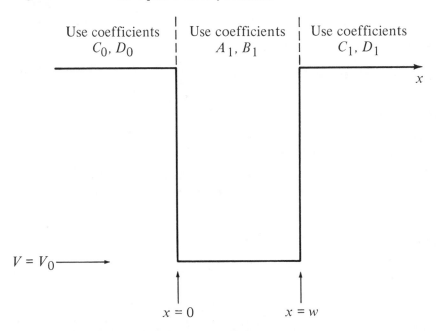

While the square well potential is never physically realized in nature, it can be used as an approximation to the nuclear potential. We study the square well potential since it illustrates many of the general properties of the wave function ψ. In addition, it is particularly simple to solve the Schrödinger equation in this case. For values of x that lie inside the well, we need to solve the equation

$$\frac{d^2\psi}{dx^2} = -\gamma_1^2 \psi, \tag{3.28}$$

where $\gamma_1 = \sqrt{E - V_0}$. (Note that γ_1 is a real number, since $V_0 < E < 0$.) As can be verified by direct substitution, the general solution inside the well is given by

$$\psi(x) = A_1 \sin \gamma_1 x + B_1 \cos \gamma_1 x, \tag{3.29}$$

where A_1 and B_1 are constants to be determined later. For values of x that lie outside the well, we need to solve the equation

$$\frac{d^2\psi}{dx^2} = \gamma_2{}^2 \, \psi, \tag{3.30}$$

where $\gamma_2 = \sqrt{-E}$. As can be verified by direct substitution, the general solution outside the well is given by

$$\psi(x) = C_j e^{\gamma_2 x} + D_j e^{-\gamma_2 x}, \tag{3.31}$$

where $j = 0$ to the left of the well and $j = 1$ to the right of the well (see Figure 3.26)

The values of the coefficients $A_1, B_1, C_0, C_1, D_0,$ and D_1 can be determined from the appropriate boundary conditions. In order that the wave function be physically meaningful, $\psi(x)$ must remain finite for $x \to \pm\infty$. (For a bound state, for which $E < 0$, $\psi(x)$ must satisfy the stronger condition $\psi(x) \to 0$ for $x \to \pm\infty$.) To satisfy this condition in the region to the left of the well, we must have $D_0 = 0$, and in the region to the right of the well, we must have $C_1 = 0$. We shall now show that these two conditions cannot be simultaneously satisfied for an arbitrary value of the energy E. For example, suppose we assume that $D_0 = 0$. We may then choose a value for C_0 arbitrarily, since only the ratio D_0/C_0 affects the shape of the wave function. Let us assume $C_0 = 1$. Using these values of C_0 and D_0, which completely defines the wave function in the region to the left of the well, we can then determine B_1 by requiring the wave function to be continuous at $x = 0$:

$$\psi(x) = A_1 \sin \gamma_1 x + B_1 \cos \gamma_1 x$$

yields $B_1 = \psi(0)$, where $\psi(0) = C_0 = 1$. Similarly, the coefficient A_1 can be found by requiring the derivative $\psi'(x)$ to be continuous at $x = 0$:

$$\psi'(x) = \gamma_1 A_1 \cos \gamma_1 x - \gamma_1 B_1 \sin \gamma_1 x$$

yields $A_1 = \dfrac{\psi'(0)}{\gamma_1}$, where $\psi'(0) = \gamma_2 C_0 = \gamma_2$. The above values for A_1 and B_1 completely define the wave function inside the well. We can therefore determine the wave function and its derivative at the right edge of the well ($x = w$) using

$$\psi(w) = A_1 \sin \gamma_1 w + B_1 \cos \gamma_1 w \tag{3.32}$$

and

$$\psi'(w) = \gamma_1 A_1 \cos \gamma_1 w - \gamma_1 B_1 \sin \gamma_1 w. \tag{3.33}$$

Continuity of the wave function and its first derivative at the right edge of the well requires that

$$\psi(w) = C_1 e^{+\gamma_2 w} + D_1 e^{-\gamma_2 w} \tag{3.34}$$

and

$$\psi'(w) = \gamma_2 C_1 e^{+\gamma_2 w} - \gamma_2 D_1 e^{-\gamma_2 w}. \tag{3.35}$$

Finally, we can solve equations 3.34 and 3.35 simultaneously to obtain C_1 and D_1:

$$C_1 = \tfrac{1}{2} e^{-\gamma_2 w} \left(\psi(w) + \frac{\psi'(w)}{\gamma_2} \right) \tag{3.36}$$

$$D_1 = \tfrac{1}{2} e^{+\gamma_2 w} \left(\psi(w) - \frac{\psi'(w)}{\gamma_2} \right),$$

where $\psi(w)$ and $\psi'(w)$ are given by equations 3.32 and 3.33. Thus, given values for C_0 and D_0, we have found values for A_1, B_1, C_1 and D_1, thereby completely defining the wave function. Since C_1 has been determined on the basis of the continuity of the wave function and its derivative, we would not, in general, expect to find $C_1 = 0$, which is the second condition for a physically meaningful wave function. (The first condition is $D_0 = 0$.) That is, physically meaningful wave functions are not obtained for *arbitrary* values of the energy E (which determines the values of γ_1 and γ_2), but only for certain values for which equation 3.36 yields $C_1 = 0$.

We illustrate this using the wave functions for three different energies, shown in Figure 3.27. All three wave functions have the proper asymptotic behavior for $x \to -\infty$ ($D_0 = 0$). However, only the wave function labeled $E = E_1$ also has the proper asymptotic behavior for $x \to +\infty$ ($C_1 = 0$). Therefore, of the three wave functions only this one is physically meaningful, i.e., an *eigenfunction*. The energy value E_1 associated with the eigenfunction is known as an *eigenvalue*, or an *energy level*. The wave function corresponding to an energy $E < E_1$ has

Figure 3.27   Wave functions for three energies including the eigenvalue E_1

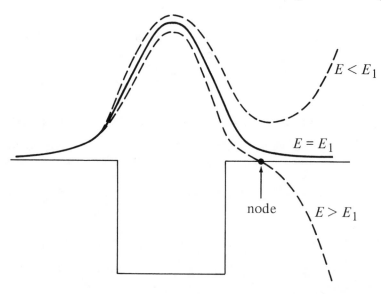

$C_1 > 0$, and therefore diverges upwards; while the one corresponding to an energy $E > E_1$ has $C_1 < 0$, and therefore diverges downwards. A second eigenvalue may be found at some higher energy, $E = E_2$. In Figure 3.28, we see the appearance of the second eigenfunction, together with wave functions corresponding to energies slightly above and below the second eigenvalue. In general, a square well has some particular number of energy levels (possibly zero), which depends on the well parameters V_0 and w. In finding solutions to the Schrödinger equation (3.27), our primary concern is to find the complete set of eigenfunctions and eigenvalues for a particular potential.

The Schrödinger equation states that the second derivative $d^2\psi/dx^2$ is equal to the function $(V(x) - E)\psi(x)$. Thus, in regions where $V(x) - E$ is positive (outside the well), the sign of the second derivative must be the same as that of the function itself, so that the wave function $\psi(x)$ curves *away from* the x-axis. Conversely, in regions where $V(x) - E$ is negative (inside the well), the sign of the second derivative must be opposite that of the function, so that the wave function $\psi(x)$ curves *towards* the x-axis. At any value of x, the magnitude of the second derivative is proportional to the wave function.

The considerations in the preceding paragraph make it easy to determine, by inspection, whether a wave function corresponds to an energy above or below a particular eigenvalue. We can illustrate the idea using the three curves shown in Figure 3.27. The curve labeled $E < E_1$ must have a smaller curvature inside the well than the eigenfunction ($E = E_1$), due to the dependence of the second derivative on the factor $V - E$. This causes the curve to have a greater height at the right edge of the well than the eigenfunction. Outside the well, its greater initial height, together with the greater magnitude of $V - E$, causes this function ($E < E_1$) to curve away from the x-axis more strongly than the eigenfunction. The greater curvature, coupled with its smaller initial slope at the right edge of the well, insures that the curve "pulls out of its dive" too soon and diverges toward $+\infty$. A similar analysis shows that the curve labeled $E > E_1$

Figure 3.28   *Wave functions for three energies including the eigenvalue E_2*

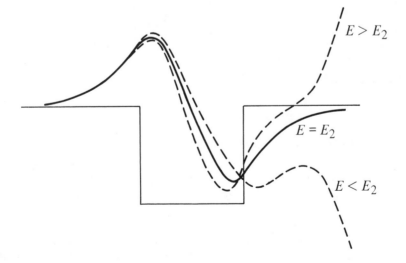

does, in fact, correspond to an energy greater than E_1. Its divergence toward $-\infty$ is due to the fact that any wave function $\psi(x)$ curves away from the x-axis to the right of the well; thus once it crosses the x-axis (with nonzero slope), it must diverge to $-\infty$.*

One other useful property of the wave function that we shall utilize extensively can also be illustrated with the aid of Figures 3.27 and 3.28. *Nodes* are points where the function crosses the x-axis. In Figure 3.27, the wave function for an energy slightly below E_1 has zero nodes while that for an energy slightly above E_1 has one node. In Figure 3.28, we see that the wave function for an energy slightly below E_2 has one node, while that for a slightly higher energy has two nodes. These observations can be generalized into the following rule: Given two solutions of the Schrödinger equation ψ_a and ψ_b (*not* necessarily eigenfunctions) which correspond to energies E_a and E_b, and which have n-1 and n nodes respectively, then the nth eigenvalue E_n must lie between E_a and E_b (E_n might exactly equal E_a). Similarly, if ψ_a and ψ_b have $n-1$ and $n+k$ nodes respectively, then the eigenvalues $E_n, E_{n+1}, \ldots, E_{n+k}$ must all lie between E_a and E_b. These two rules apply to any type of potential. We shall use them to determine empirically, with the aid of a computer, all the eigenvalues and eigenfunctions for single and multiple square well potentials.

The Multiple Square Well Potential

The form of the solution of the Schrödinger equation for a multiple square well potential (see Figure 3.29) is the same as that for the single square well potential. Inside each well, the solution is given by equation 3.29; outside each

Figure 3.29   First three wells of a multiple well potential

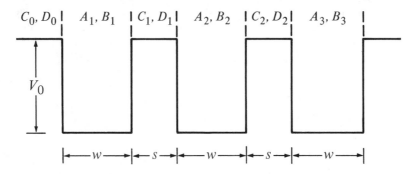

For the kth well the wave function is

$$\psi = A_k \sin \gamma_1 x + B_k \cos \gamma_1 x$$
or
$$\psi = C_k e^{+\gamma_2 x} + D_k e^{-\gamma_2 x}.$$

---

*   The same arguments presented here can be applied to a well of *any* shape. For wells other than the square well, the value of x at which the wave function changes the sign of its curvature is not a fixed value (the edge of the well) but rather a function of the energy E, i.e., the value of x that satisfies $V(x) - E = 0$.

well, the solution is given by equation 3.31. As in the case of the single well, we choose the coefficients $C_0 = 1$ and $D_0 = 0$, so that the wave function has the proper asymptotic form for $x \to -\infty$. In each region, starting with the first well, and going from left to right, we can determine the coefficients A_k and B_k (or C_k and D_k) from the coefficients in the previous region, by requiring the wave function and its slope to be continuous at each boundary, as for a single well. The wave functions determined from this procedure only have the proper asymptotic behavior for $x \to +\infty$ if the energy E happens to be an eigenvalue. As discussed above, we can use the rules concerning the number of nodes in the wave function to empirically locate all the eigenvalues and eigenfunctions for a multiple well potential.

The eigenvalues and eigenfunctions for the multiple well potential bear an important relationship to those of the single well. To illustrate the connection, let us consider the double well potential. We assume that the depth V_0 and the width w of each well are the same as those for the single well considered previously. The first two eigenfunctions for the double well are shown in Figures 3.30 and 3.31. We can easily show that the eigenvalues associated with the first and second eigenfunctions are, respectively, slightly below and slightly above the first single well eigenvalue E_1. Only a small increase in the curvature of the first double well eigenfunction would be needed for it to coincide with the single well eigenfunction (shown dotted in Figures 3.30 and 3.31). Therefore, the first double well eigenvalue is slightly less than the first single well eigenvalue, according to the relationship between energy and the curvature of the wave function. A similar argument can be applied in the case of the second double well eigenfunction to demonstrate that the second eigenvalue for the double well is slightly greater than the first single well eigenvalue. The other eigenfunctions and eigenvalues for the double well are related to those of the single well in an analogous manner. For each single well eigenvalue, there are two double well eigenvalues and two associated eigenfunctions. One double well eigenvalue lies below the single well eigenvalue, and corresponds to a symmetric wave function; the other lies above the single well eigenvalue and corresponds to an anti-symmetric wave function.

If we extend the above observations to the case of a potential consisting of n wells, we find that we have a set of n closely spaced eigenvalues for each single

Figure 3.30   First double well eigenfunction

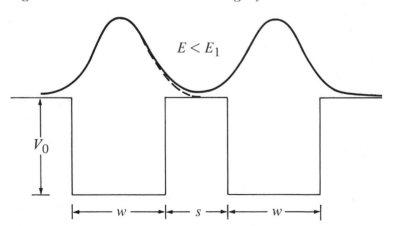

Figure 3.31   Second double well eigenfunction

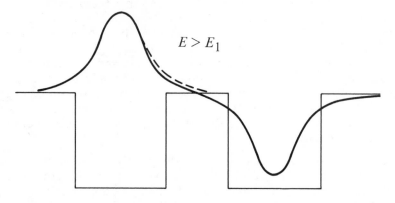

well eigenvalue. In this case, however, the associated eigenfunctions are not necessarily symmetric or antisymmetric. If we allow the number of wells n to become infinite, we have the case of a periodic potential. A periodic potential is physically realized for an electron moving through a crystalline solid, although in the real case the shape of each well is not square and the problem is not one-dimensional. Based on the above statements concerning the n-well potential, we would expect to find in the case of a periodic potential, an infinite number of closely spaced eigenvalues, or an *energy band*, corresponding to each single well eigenvalue. The existence of such energy bands in crystalline solids is central to solid state physical theory and has been amply verified experimentally.

Computer Program to Determine the Eigenvalues and Eigenfunctions

The computer program shown in flow diagram form in Figures 3.41-3.43 and listed on pages 164-168, can be used to find all the eigenvalues and eigenfunctions for specified single and multiple square well potentials up to 20 wells. The program first reads from a data card, numerical values for the parameters V, W, NW, S, EMIN, EMAX, and NE, where

$V = V_0$ (depth of each potential well)
$W = w$ (width of each potential well)
$NW = n$ (number of wells)
S = separation between wells
$EMIN$ = smallest energy to be used in seeking eigenvalues
$EMAX$ = largest energy to be used in seeking eigenvalues
NE = number of energies to be used in the interval (EMIN, EMAX).

After reading the data card, the program sets some parameters for later use:

S = separation between adjacent wells, is set to the well width in the event there is only a single well;
Y = a distance which includes all NW wells;
ΔE = spacing between trial energy values, is set to zero if there is only one energy (NE = 1), otherwise it is set to $(EMAX - EMIN)/(NE - 1)$;
E = the first trial energy, is set to EMIN.

The program then enters a loop over NE + 1 energy values. For *each* energy E the program calculates the wave function coefficients A_k, B_k, C_k, and D_k for *each* well, $k = 1, 2, \ldots, NW$, by requiring the wave function and its derivative be continuous at each well boundary. The program also calculates and prints out the number of nodes in the wave function for each of the first NE energy values. (The reason that the loop over energy values includes NE + 1 energies—one more than the specified number—is that on the last time through the loop, the program computes the wave function coefficients A_k, B_k, C_k, and D_k, using the average energy $\frac{1}{2}$ (EMIN + EMAX). These are used later to make a plot of the wave function having this energy.

The technique used to find the number of nodes in the wave function will now be described.* Inside one of the NW wells, suppose the wave function is as shown in Figure 3.32. The shape of the sine and cosine functions is such that if the well contains between N and N + 1 half-wavelengths, then the number of nodes inside the well is either N or N + 1 according to the following scheme:

number of nodes in well = N   provided N is even and ψ has opposite signs at the two edges of the well ($\psi_1 \psi_2 < 0$)

                                        or

                                       N is odd and ψ has the same signs at the two edges of the well ($\psi_1 \psi_2 > 0$)

number of nodes in well = N + 1   otherwise.

Figure 3.32   Possible wave function

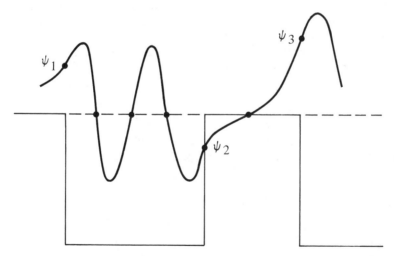

---

* A technique that is simpler than the one described here, which could be used for potential wells of any shape, is to first calculate the wave function at a series of x-values $x_1, x_2, x_3, \ldots, x_N$. All the nodes between x_1 and x_N can then be counted by going through the series $\psi_1, \psi_2, \psi_3, \ldots, \psi_N$ and adding one to the number of nodes each time consecutive terms change sign. However, this simple technique may miss some nodes for rapidly varying wave functions having many nodes.

In the region between wells shown in Figure 3.32, the wave function has one additional node provided it changes sign, i.e., $\psi_2 \psi_3 < 0$. After the loop over the number of wells is complete, the variable NODES contains the number of nodes in the wave function in all NW wells. However, there may be yet one more node beyond the last (rightmost) well. This can be determined by using the wave function coefficients for this last region, C_{NW} and D_{NW}, to find the point where the function $C_{NW} e^{+\gamma_2 X} + D_{NW} e^{-\gamma_2 X}$ crosses the x-axis. If the location of this point (x_0) is in the region beyond the last well, then the variable NODES is increased by one.

Having calculated the number of nodes in the entire wave function, the program prints out the trial energy E and the number of nodes, then increments E by ΔE to repeat the whole process for the next energy value. For the last time through the loop over NE + 1 energies, the average value $\frac{1}{2}$ (EMIN + EMAX) is used. In this case, there is no printout of E and NODES. Instead, the program produces a graph for this particular wave function, together with a graph of the potential. The plotted wave function is not, in general, an eigenfunction, unless the energy $E = \frac{1}{2}$ (EMIN + EMAX) happens to be an eigenvalue. Thus, while the wave function has the proper asymptotic behavior for $x \to -\infty$, it generally diverges for $x \to +\infty$.

In order to produce the plot, the program first goes through a loop over 51 x-values covering the entire region of all NW wells. For each of the 51 x-values, the program determines which well (number k) this x-value corresponds to, and uses the previously determined coefficients A_k, B_k (if x is *inside* the kth well) or C_k, D_k (if x is *between* wells) to compute the wave function according to equation 3.29 or 3.31. In the same loop, the program computes a scaled value of the potential at each x-value to use in making the plot. It uses U = 51 to represent V = 0, the potential between wells, and U = 10 to represent $V = V_0$, the potential inside a well. In addition, the program determines the largest absolute value of the wave function ψ_{max} at any of the 51 x-values except those to the right of the last well (where the wave function may diverge), and later uses ψ_{max} to scale the plot.

The program continues in Figure 3.42, where it enters a final loop over 51 x-values $x_1, x_2, \ldots, x_{51}$, at which the wave function has been previously computed. To make the plot, it uses the linear scaling function

$$M1 = 25 \frac{\psi_j}{\psi_{max}} + 51.5.$$

This scaling function is chosen so that for $|\psi_j| \leq \psi_{max}$ M1 is in the range $26 \leq M1 \leq 76$, with M1 = 51 for $\psi_j = 0$. The program then checks that M1 is inside the allowed range $1 \leq M1 \leq 101$. It stores dots ('.') in the appropriate location of the array Z (either Z_{10} or Z_{51}), to represent the potential. At a step in the potential it stores dots in $Z_{10}, Z_{11}, \ldots, Z_{51}$. Finally, it stores an asterisk ('*') in Z_{M1} to represent the potential, and prints out on a single line, the numerical values of x_j and ψ_j, together with the array of characters Z_1, $Z_2, \ldots, Z_{101}$. It then proceeds to the next x-value, printing the plot one line at a time, until all 51 lines are complete. Then the program proceeds to read the next data card, if any remain to be read.

Comments on Data Cards and Sample Results

The output shown in Figures 3.33-3.34 can be produced using a data card containing the following numerical values for the parameters:

V	W	NW	S	EMIN	EMAX	NE
−20.	1.0	1.0	1.0	−20.	0.	51.

In this case, the value used for the parameter S does not matter, since the data card specifies a single square well potential (NW = 1) of depth $V_0 = -20$. The reason for specifying EMIN = −20., and EMAX = 0.0, is that we wish to cover the entire range of possible energies, since we have no information on the eigenvalues. The value used for NE (51.) is arbitrary, and was chosen to obtain a convenient spacing between each of the 51 energy values (0.4).

Figure 3.33

```
V= -20.00000  W=   1.00000 NW=   1.00000  S=   1.00000
EMIN= -20.00000  EMAX=    .00000 NE=  51.00000
```

E	NODES		E	NODES
−20.00000	0		−9.60000	1
−19.60000	0		−9.20000	1
−19.20000	0		−8.80000	1
−18.80000	0		−8.40000	1
−18.40000	0		−8.00000	1
−18.00000	0		−7.60000	1
−17.60000	0		−7.20000	1
−17.20000	0		−6.80000	1
−16.80000	0		−6.40000	1
−16.40000	0		−6.00000	1
−16.00000	0		−5.60000	1
−15.60000	0		−5.20000	1
−15.20000	1		−4.80000	1
−14.80000	1		−4.40000	1
−14.40000	1		−4.00000	1
−14.00000	1		−3.60000	2
−13.60000	1		−3.20000	2
−13.20000	1		−2.80000	2
−12.80000	1		−2.40000	2
−12.40000	1		−2.00000	2
−12.00000	1		−1.60000	2
−11.60000	1		−1.20000	2
−11.20000	1		−.80000	2
−10.80000	1		−.40000	2
−10.40000	1		.00000	1
−10.00000	1			

According to the output in Figure 3.33, the number of nodes in the wave function changes from zero to one, between E = −15.6 and E = −15.2. The number of nodes also changes from one to two between E = −4.0 and E = −3.6. These results imply that there are *two* eigenvalues E_1 and E_2 for this particular single well potential, and that E_1 lies between −15.6 and −15.2, and E_2 lies between −4.0 and −3.6. In other words, the eigenvalues are

$$E_1 = -15.4 \pm 0.2$$
$$E_2 = -3.8 \pm 0.2.$$

Figure 3.34

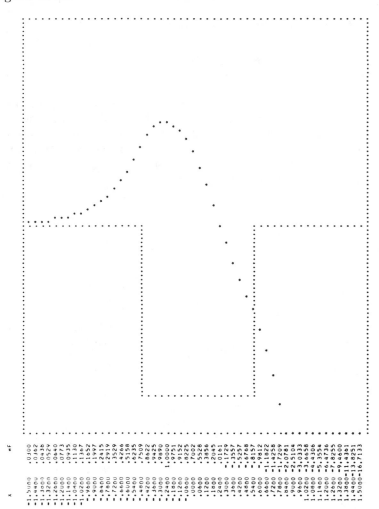

The wave function shown plotted in Figure 3.34 for the energy $E = -10.0$ (the average of -20.0 and 0.0) is, of course, far from being an eigenfunction.

Having obtained approximate values for the eigenvalues, we can run the program again to obtain more accurate values by using values for EMIN and EMAX that correspond to the known limits on each eigenvalue. For example, the output shown in Figures 3.35-3.36 was obtained using a data card containing the following numerical values for the parameters:

V	W	NW	S	EMIN	EMAX	NE
−20.	1.0	1.0	1.0	−15.6	−15.2	51.

Based on the output in Figure 3.35, we see that the number of nodes in the wave function changes from zero to one between $E = -15.416$ and $E = -15.408$, which means that the first eigenvalue E_1 is given by

$$E_1 = -15.412 \pm .004.$$

Note that the uncertainty in the value of E_1 has been reduced by a factor of fifty from the first run. The wave function shown in Figure 3.36 for the energy $E = -15.4$ (the average of -15.6 and -15.2) is much closer to being an eigenfunction than that in Figure 3.34, for $E = -10$. By repeating this process of using a successively narrower energy range EMIN to EMAX, we can determine each eigenvalue and eigenfunction to any desired accuracy.

Figure 3.35

```
V= -20.00000   W=   1.00000  NW=   1.00000  S=   1.00000
EMIN= -15.60000   EMAX= -15.20000  NE=   51.00000


      E              NODES


  -15.60000            0                    -15.39200          1
  -15.59200            0                    -15.38400          1
  -15.58400            0                    -15.37600          1
  -15.57600            0                    -15.36800          1
  -15.56800            0                    -15.36000          1
  -15.56000            0                    -15.35200          1
  -15.55200            0                    -15.34400          1
  -15.54400            0                    -15.33600          1
  -15.53600            0                    -15.32800          1
  -15.52800            0                    -15.32000          1
  -15.52000            0                    -15.31200          1
  -15.51200            0                    -15.30400          1
  -15.50400            0                    -15.29600          1
  -15.49600            0                    -15.28800          1
  -15.48800            0                    -15.28000          1
  -15.48000            0                    -15.27200          1
  -15.47200            0                    -15.26400          1
  -15.46400            0                    -15.25600          1
  -15.45600            0                    -15.24800          1
  -15.44800            0                    -15.24000          1
  -15.44000            0                    -15.23200          1
  -15.43200            0                    -15.22400          1
  -15.42400            0                    -15.21600          1
  -15.41600            0                    -15.20800          1
  -15.40800            1                    -15.20000          1
  -15.40000            1
```

The procedure for finding the eigenvalues and eigenfunctions for a multiple square well potential is exactly the same as that just described for a single square well. For example, the output shown in Figures 3.37-3.40 was obtained using the following values for the parameters:

	V	W	NW	S	EMIN	EMAX	NE
1st card	−20.	1.0	2.0	1.0	−15.464	−15.456	51.
2nd card	−20.	1.0	2.0	1.0	−15.368	−15.360	51.

As noted previously, the two lowest eigenvalues for the double well lie below and above the lowest single well eigenvalue. The first double well eigenfunction is symmetric (Figure 3.38), and the second is antisymmetric (Figure 3.40).

Figure 3.36

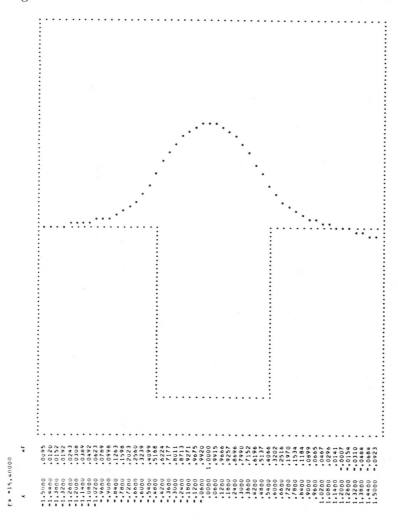

Problems 20-25 at the end of this chapter pertain to the solution of the Schrödinger equation.

Figure 3.37

```
V= -20.00000  W=  1.00000 NW=  2.00000 S=  1.00000
EMIN= -15.46400  EMAX= -15.45600 NE=  51.00000
```

E	NODES				
-15.46400	0	-15.46128	0	-15.45856	0
-15.46384	0	-15.46112	0	-15.45840	0
-15.46368	0	-15.46096	0	-15.45824	0
-15.46352	0	-15.46080	0	-15.45808	0
-15.46336	0	-15.46064	0	-15.45792	0
-15.46320	0	-15.46048	0	-15.45776	0
-15.46304	0	-15.46032	0	-15.45760	0
-15.46288	0	-15.46016	0	-15.45744	0
-15.46272	0	-15.46000	0	-15.45728	0
-15.46256	0	-15.45984	0	-15.45712	0
-15.46240	0	-15.45968	0	-15.45696	1
-15.46224	0	-15.45952	0	-15.45680	1
-15.46208	0	-15.45936	0	-15.45664	1
-15.46192	0	-15.45920	0	-15.45648	1
-15.46176	0	-15.45904	0	-15.45632	1
-15.46160	0	-15.45888	0	-15.45616	1
-15.46144	0	-15.45872	0	-15.45600	1

Figure 3.38

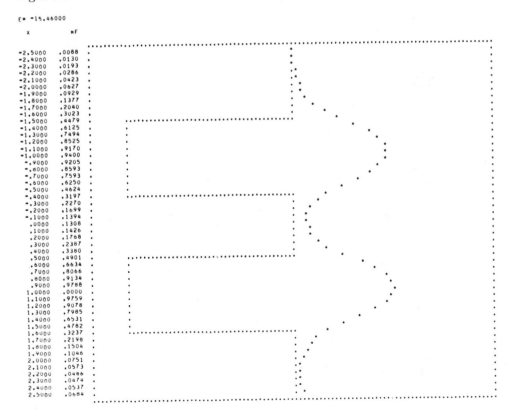

```
E= -15.46000

    X        WF
 -2.5000   .0088
 -2.4000   .0130
 -2.3000   .0193
 -2.2000   .0286
 -2.1000   .0423
 -2.0000   .0627
 -1.9000   .0929
 -1.8000   .1377
 -1.7000   .2040
 -1.6000   .3023
 -1.5000   .4479
 -1.4000   .6125
 -1.3000   .7494
 -1.2000   .8525
 -1.1000   .9170
 -1.0000   .9400
  -.9000   .9205
  -.8000   .8593
  -.7000   .7593
  -.6000   .6250
  -.5000   .4624
  -.4000   .3197
  -.3000   .2270
  -.2000   .1699
  -.1000   .1394
   .0000   .1308
   .1000   .1426
   .2000   .1768
   .3000   .2387
   .4000   .3380
   .5000   .4901
   .6000   .6634
   .7000   .8066
   .8000   .9134
   .9000   .9788
  1.0000   .0000
  1.1000   .9759
  1.2000   .9078
  1.3000   .7985
  1.4000   .6531
  1.5000   .4782
  1.6000   .3237
  1.7000   .2198
  1.8000   .1504
  1.9000   .1046
  2.0000   .0751
  2.1000   .0573
  2.2000   .0486
  2.3000   .0474
  2.4000   .0537
  2.5000   .0684
```

Figure 3.39

```
V= -20.00000  W=   1.00000 NW=   2.00000  S=   1.00000
EMIN= -15.36800  EMAX= -15.36000  NE=  51.00000
```

E	NODES	E	NODES	E	NODES
-15.36800	1	-15.36528	1	-15.36256	2
-15.36784	1	-15.36512	1	-15.36240	2
-15.36768	1	-15.36496	1	-15.36224	2
-15.36752	1	-15.36480	1	-15.36208	2
-15.36736	1	-15.36464	1	-15.36192	2
-15.36720	1	-15.36448	1	-15.36176	2
-15.36704	1	-15.36432	1	-15.36160	2
-15.36688	1	-15.36416	1	-15.36144	2
-15.36672	1	-15.36400	1	-15.36128	2
-15.36656	1	-15.36384	1	-15.36112	2
-15.36640	1	-15.36368	1	-15.36096	2
-15.36624	1	-15.36352	1	-15.36080	2
-15.36608	1	-15.36336	1	-15.36064	2
-15.36592	1	-15.36320	1	-15.36048	2
-15.36576	1	-15.36304	1	-15.36032	2
-15.36560	1	-15.36288	1	-15.36016	2
-15.36544	1	-15.36272	2	-15.36000	2

Figure 3.40

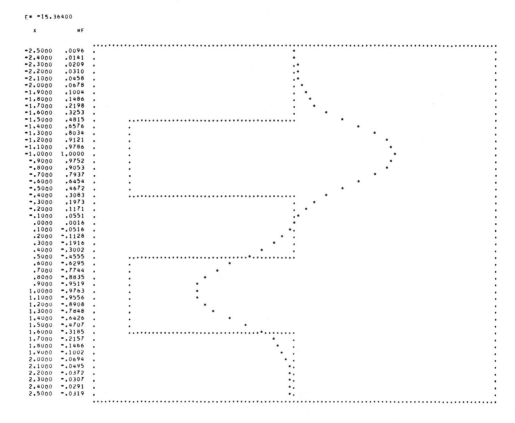

```
E= -15.36400

   X        WF
-2.5000   .0096
-2.4000   .0141
-2.3000   .0209
-2.2000   .0310
-2.1000   .0458
-2.0000   .0678
-1.9000   .1004
-1.8000   .1486
-1.7000   .2198
-1.6000   .3253
-1.5000   .4815
-1.4000   .6576
-1.3000   .8034
-1.2000   .9121
-1.1000   .9786
-1.0000  1.0000
 -.9000   .9752
 -.8000   .9053
 -.7000   .7937
 -.6000   .6454
 -.5000   .4672
 -.4000   .3083
 -.3000   .1973
 -.2000   .1171
 -.1000   .0551
  .0000   .0016
  .1000  -.0516
  .2000  -.1128
  .3000  -.1916
  .4000  -.3002
  .5000  -.4555
  .6000  -.6295
  .7000  -.7744
  .8000  -.8835
  .9000  -.9519
 1.0000  -.9763
 1.1000  -.9556
 1.2000  -.8908
 1.3000  -.7848
 1.4000  -.6426
 1.5000  -.4707
 1.6000  -.3185
 1.7000  -.2157
 1.8000  -.1466
 1.9000  -.1002
 2.0000  -.0694
 2.1000  -.0495
 2.2000  -.0372
 2.3000  -.0307
 2.4000  -.0291
 2.5000  -.0319
```

Figure 3.41   Flow diagram to solve the Schrödinger equation—Part 1

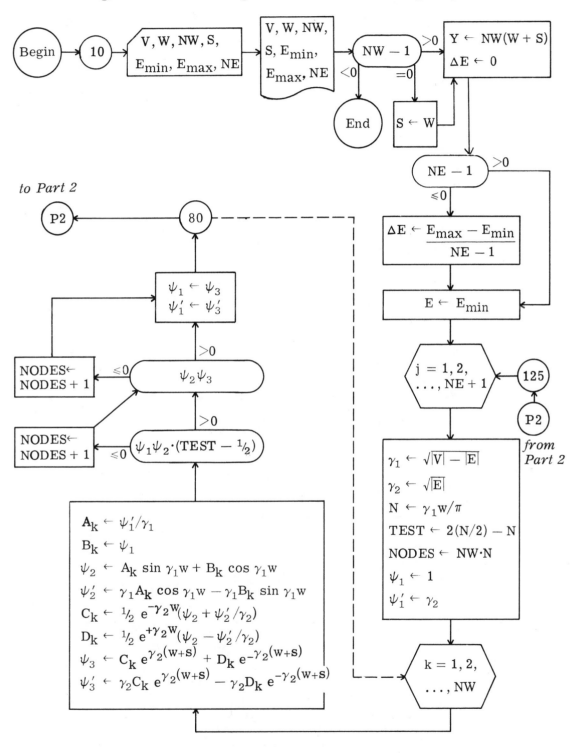

Figure 3.42   *Flow diagram to solve the Schrödinger equation—Part 2*

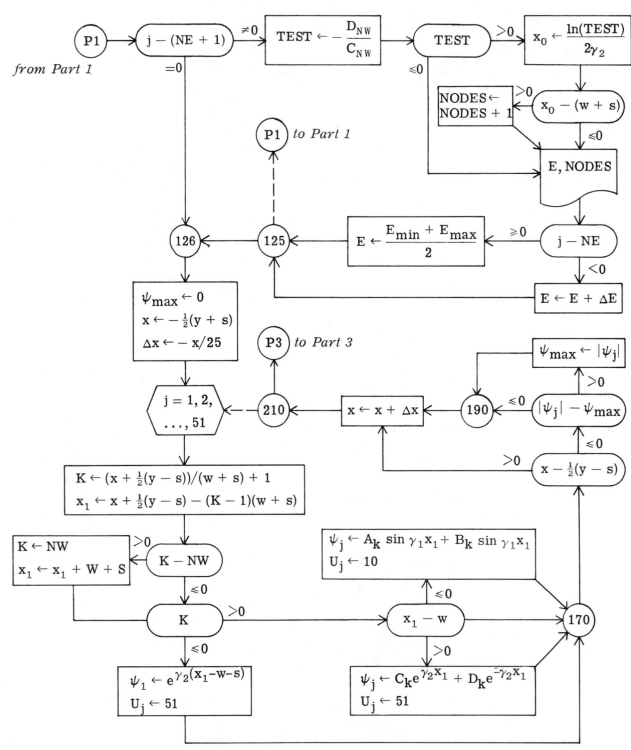

Figure 3.43   Flow diagram to solve the Schrödinger equation—Part 3

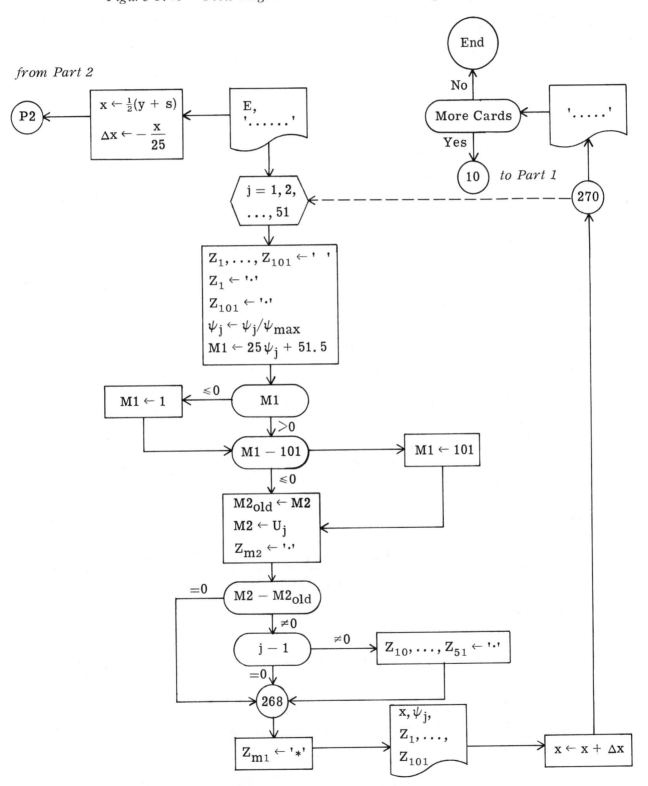

```
C                      SOLUTION OF THE SCHRODINGER EQUATION FOR MULTIPLE SQUARE WELLS
C
C--------------------------------------------------------------------------------
C                      PROGRAM FINDS THE NUMBER OF NODES IN THE WAVE FUNCTION FOR
C                      A SERIES OF TRIAL ENERGY VALUES BETWEEN EMIN AND EMAX.
C                      IT ALSO DISPLAYS A PLOT OF THE WAVE FUNCTION FOR THE AVERAGE
C                      ENERGY (EMIN+EMAX)/2.0
C
C
      REAL NW,NE
      DATA BLANK,ASTER,DOT/' ','*','.'/
      DIMENSION Z(101),WF(51),U(51),A(20),B(20),C(20),D(20)
      PI=3.14159265
C
C                      READ VALUES FOR V, W, NW, S, EMIN, EMAX, NE
C                      V=WELL DEPTH
C                      W=WELL WIDTH
C                      NW=NUMBER OF WELLS
C                      S=SEPARATION OF WELLS
C                      EMIN=MINIMUM TRIAL ENERGY
C                      EMAX =MAXIMUM TRIAL ENERGY
C                      NE=NUMBER OF TRIAL ENERGIES
C
   10 READ(2,1000)V,W,NW,S,EMIN,EMAX,NE
      WRITE(3,1001)V,W,NW,S,EMIN,EMAX,NE
      IF(NW-1.0)280,20,30
C
C                 FOR ONE WELL, ASSUME S=W
C
   20 S=W
   30 Y=NW*(W+S)
      MW=NW
      ME=NE
      DE=0.
      IF(ME-1)34,34,32
C
C              DE IS THE SPACING BETWEEN ADJACENT TRIAL ENERGY VALUES.
C
   32 DE=(EMAX-EMIN)/(NE-1.0)
   34 CONTINUE
      E=EMIN
      ME1=ME+1
C
C              LOOP OVER NE+1 ENERGY VALUES.  (THE ONE EXTRA TIME IS TO TREAT THE
C              CASE  E=(EMIN+EMAX)/2.0.  THE PURPOSE OF THE LOOP IS TO COMPUTE
C              THE NUMBER OF NODES AND THE COEFFICIENTS A, B, C, D.
C
      DO 125 J=1,ME1
      G1=SQRT(ABS(ABS(V)-ABS(E)))
      G2=SQRT(ABS(E))
C
C              N IS THE NUMBER OF HALF-WAVELENGTHS PER WELL.
C
      N=G1*W/PI
      TEST=2*(N/2)-N
C
C              NODES IS THE NUMBER OF NODES IN ALL WELLS.
C
      NODES=MW*N
C
C              SET INITIAL VALUES OF WAVE FUNCTION AND ITS DERIVATIVE AT THE
```

```
C             EDGE OF THE FIRST WELL.
C
      WF1=1.0
      DWF1=G2
C
C             LOOP OVER THE NUMBER OF WELLS.
C
      DO 80 K=1,MW
C
C             FOR EACH WELL NUMBER K, COMPUTE THE WAVE FUNCTION COEFFICIENTS
C             A(K), B(K),C(K), D(K), BY REQUIRING CONTINUITY OF THE WAVE
C             FUNCTION (WF)  AND ITS DERIVATIVE  (DWF).
C
      A(K)=DWF1/G1
      B(K)=WF1
C
C             FIND WAVE FUNCTION AND DERIVATIVE AT NEXT BOUNDARY.
C
      WF2=A(K)*SIN(G1*W)+B(K)*COS(G1*W)
      DWF2=G1*(A(K)*COS(G1*W)-B(K)*SIN(G1*W))
      C(K)=0.5*EXP(-G2*W)*(WF2+DWF2/G2)
      D(K)=0.5*EXP(+G2*W)*(WF2-DWF2/G2)
C
C             FIND WAVE FUNCTION AND DERIVATIVE AT START OF NEXT WELL.
C
      WF3=C(K)*EXP(+G2*(W+S))+D(K)*EXP(-G2*(W+S))
      DWF3=G2*(C(K)*EXP(+G2*(W+S))-D(K)*EXP(-G2*(W+S)))
C
C             THE VALUE OF NODES IS CORRECTED ACCORDING TO THE RELATIVE SIGNS
C             OF WF1 AND WF2 AND WHETHER THE NUMBER OF HALF-WAVELENGTHS IN
C             THE WELL IS EVEN OR ODD.
C
      IF(WF1*WF2*(TEST+0.5))40,40,50
   40 NODES=NODES+1
   50 IF(WF2*WF3)60,60,70
   60 NODES=NODES+1
C
C             DO NEXT WELL.
C             K IS THE WELL NUMBER.
C
      K=(X+(Y-S)/2.0)/(W+S)+1.0
      XK=K
C
C             X1 IS THE X COORDINATE RELATIVE TO THE EDGE OF THE KTH WELL.
C
      X1=X+(Y-S)/2.0-(XK-1.0)*(W+S)
      IF(K-MW)128,128,127
  127 K=MW
      X1=X1+(W+S)
C
C             COMPUTE WF(J) AND U(J) ACCORDING TO A FORMULA WHICH DEPENDS ON
C             K AND X1.
C
  128 IF(K)130,130,140
  130 WF(J)=EXP(+G2*(X1-W-S))
      U(J)=51.
      GO TO 170
  140 IF(X1-W)150,150,160
  150 WF(J)=A(K)*SIN(G1*X1)+B(K)*COS(G1*X1)
      U(J)=10.
      GO TO 170
  160 WF(J)=C(K)*EXP(+G2*X1)+D(K)*EXP(-G2*X1)
      U(J)=51.
```

```
  170 IF(X-(Y-S)/2.0)175,175,210
C
C
C              COMPUTE THE LARGEST MAGNITUDE OF THE WAVE FUNCTION, (WFMAX),
C              FOR SCALING PURPOSES IN MAKING PLOT.
C
  175 IF(ABS(WF(J))-WFMAX)190,190,180
  180 WFMAX=ABS(WF(J))
  190 CONTINUE
C
C              DO NEXT X-VALUE.
C
  210 X=X+DX
C
C              SET INITIAL X-VALUE AND STEP SIZE FOR PLOT.
C
      X=-(Y+S)/2.0
      DX=-X/25.0
      WRITE(3,1005)E
      WRITE(3,1004)
C
C              LOOP OVER 51 X-VALUES TO MAKE PLOT.
C
      DO 270 J=1,51
C
C              SET ALL 101 POSITIONS ON A LINE TO BLANKS INITIALLY.
C
      DO 220 K=1,101
  220 Z(K)=BLANK
      Z(1)=DOT
      Z(101)=DOT
      WF(J)=WF(J)/WFMAX
C
C              COMPUTE INDEX M1, BASED ON VALUE OF THE WAVE FUNCTION.
C
      M1=25.0*WF(J)+51.5
C
C              BE SURE M1 IS BETWEEN 1 AND 101.
C
      IF(M1)230,230,240
  230 M1=1
  240 IF(M1-101)260,260,250
  250 M1=101
C
C              M2 IS COMPUTED FROM THE POTENTIAL U(J).  M2OLD IS THE PREVIOUS
C              VALUE OF M2.
C
  260 M2OLD=M2
      M2=U(J)
      Z(M2)=DOT
C
C              IF THERE IS A STEP IN THE POTENTIAL AT THIS POINT, SET Z(10),
C              Z(11),...., Z(51) = DOT.
C
      IF(M2-M2OLD)262,268,262
  262 IF(J-1)264,268,264
  264 DO 266 K=10,51
  266 Z(K)=DOT
  268 CONTINUE
C
C              STORE AN ASTERISK IN Z(M1) TO REPRESENT THE VALUE OF THE WAVE
C              FUNCTION.
C
      Z(M1)=ASTER
```

```
      WRITE(3,1003)X,WF(J),(Z(K),K=1,101)
C
C                DO NEXT X-VALUE.
C
  270 X=X+DX
      WRITE(3,1004)
C
C                READ NEXT DATA CARD.
C
      DO 270 J=1,51
C
C                SET ALL 101 POSITIONS ON A LINE TO BLANKS INITIALLY.
C
      DO 220 K=1,101
  220 Z(K)=BLANK
      Z(1)=DOT
      Z(101)=DOT
      WF(J)=WF(J)/WFMAX
C
C                COMPUTE INDEX M1, BASED ON VALUE OF THE WAVE FUNCTION.
C
      M1=25.0*WF(J)+51.5
C
C                BE SURE M1 IS BETWEEN 1 AND 101.
C
      IF(M1)230,230,240
  230 M1=1
  240 IF(M1-101)260,260,250
  250 M1=101
C
C                M2 IS COMPUTED FROM THE POTENTIAL U(J).   M2OLD IS THE PREVIOUS
C                VALUE OF M2.
C
  260 M2OLD=M2
      M2=U(J)
      Z(M2)=DOT
C
C                IF THERE IS A STEP IN THE POTENTIAL AT THIS POINT, SET Z(10),
C                Z(11),..., Z(51) = DOT.
C
      IF(M2-M2OLD)262,268,262
  262 IF(J-1)264,268,264
  264 DO 266 K=10,51
  266 Z(K)=DOT
  268 CONTINUE
C
C                STORE AN ASTERISK IN Z(M1) TO REPRESENT THE VALUE OF THE WAVE
C                FUNCTION.
C
      Z(M1)=ASTER
      WRITE(3,1003)X,WF(J),(Z(K),K=1,101)
C
C                DO NEXT X-VALUE.
C
  270 X=X+DX
      WRITE(3,1004)
C
C                READ NEXT DATA CARD.
C
      GO TO 10
 1000 FORMAT(7F10.5)
 1001 FORMAT(4H1 V=,F10.5,4H  W=,F10.5,4H NW=,F10.5,4H  S=,F10.5,/6H EMI
```

```
      1N=,F10.5,7H  EMAX=,F10.5,4H NE=,F10.5//22H      E              NODES/)
1002 FORMAT(1X,F10.5,I8)
1003 FORMAT(1X,2F8.4,2X,101A1)
1004 FORMAT(19X,101H....................................................................
    1.............................................................................)
1005 FORMAT(4H1 E=,F10.5//17H      X              WF/)
 280 CONTINUE
     CALL EXIT
     END
```

Problems for Chapter 3

Fourier Synthesis Program

1. Dependence on N, the number of waves.   Discuss the results you obtain when running the program using a series of N values between 1 and 100 for each of the four wave forms. (N = 48 is the maximum value for the female profile wave form—FLAG = 3.) An interesting phenomenon occurs for the case N = 50. Try using N = 40, 50, and 60 for the case of the square wave (FLAG = 0), and see if you can explain the differences. Discuss both the goodness of the approximation as a function of N for each wave form separately and how the goodness of the approximation depends on the nature of the wave form, for particular values of N.

2. Average deviation plots.   Compare the average deviation plots for each wave form (obtained using N = 100), and discuss the dependence of the average deviation Δs_N on N in each case. Try to relate your observations to the formulas for the coefficients a_k and b_k (equations 3.7-3.9). Superimposed on a smooth variation with N, the average deviation plots exhibit a number of bumps and dips at certain values of N. Try to find the cause of these irregularities, which are particularly prominent in the case of the spiked wave form.

3. Width of the peak for the spiked wave form.   Run the program for the spiked wave form (FLAG = 3.0) using various values for N, the number of waves. For each graph, determine (from the plot) the full-width of the peak Δx, and compare your result with the full-width predicted according to equation 3.12. Unless you expand the scale of the plot (see problem 4), this can only be done for values of N that are not too large.

4. Expansion of scale.   Examination of the detailed shape of each Fourier synthesized wave form in the vicinity of the origin is desirable, since we anticipate the approximation to be poorest at this point because of the discontinuities. The printer-generated plots do not show fine details very well, due to their low resolution. Therefore, it is useful to expand the scale of the plot near the origin, particularly for the case of a large number of waves, which give the finest detail. This can be accomplished with a minor program modification. In the present version of the program, the initial x-value and the step size Δx are assigned the values $-\pi$ and $2\pi/50$, respectively. If instead, they are assigned the values $-\pi/N$ and $2\pi/(50N)$, this will have the effect of expanding the scale in the vicinity of the origin by an amount which is proportional to the number of waves N. Discuss the results you obtain after you make this modification, and run the program for each wave form using various values for N.

5. Saw tooth wave form.   Modify the program, so that it can Fourier synthesize the "saw tooth" wave form depicted in Figure 3.44. The Fourier coefficients are given by

$$a_k = 0 \qquad\qquad b_k = \frac{2}{\pi k} \, (-1)^{k+1} \, .$$

Compare the rate at which the approximation improves with N, with the other wave forms.

Figure 3.44   The saw tooth wave form

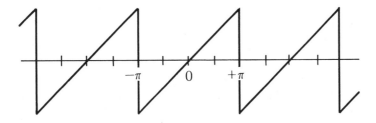

6. Square pulse wave form.   Modify the program so that it can Fourier synthesize the "square pulse" wave form depicted in Figure 3.45. The Fourier coefficients are given by

$$a_k = \frac{2}{\pi k} \, \sin k x_0 \qquad\qquad b_k = 0.$$

Run the program using various values for the parameter x_0, which is the *half-width* of the pulse (see Figure 3.45). For each value of x_0, use a number of values for N, the number of waves. Discuss how the number of waves required to get a reasonably good approximation to the wave form depends on x_0. A graph of the average deviations for $k = 2, 4, 6, \ldots, 100$ waves, which is printed when $N = 100$, would be useful for this discussion.

For a particular value of x_0, say $x_0 = .05$, make a plot (by hand or computer) of a_k versus k, for $k = 1, 2, \ldots, 25$, and discuss the shape of this graph.

Figure 3.45   The square pulse wave form

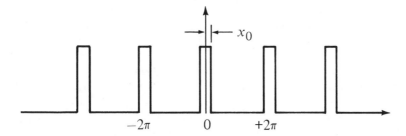

7. Other wave forms.   The approach we have followed until now has been to determine appropriate values for the coefficients $a_k, b_k, k = 1, 2, \ldots, N$, for each specific wave form. An alternative approach is to simply try different formulas for the coefficients and see what wave forms result.

Modify the program so that the coefficients $a_k, b_k, k = 1, 2, \ldots, N$, are calculated using various "cooked-up" formulas. Run the modified version of the program for several values of N, and see what wave forms appear. There are several points you should bear in mind: (1) The formulas for the coefficients should be such that $a_1 > a_2 > a_3 > \ldots > a_N$, and $b_1 > b_2 > b_3 > \ldots > b_N$, otherwise the results will (probably) not converge as N gets large.* (2) For arbitrary choices of these coefficients, the function $F_N(x)$ may go off scale when it is being plotted. It is, therefore, essential that after the program computes the integer M from the scaling function $M = 25.(S + 2.06)$, you add several statements to the program to insure that M stays within the allowed range $1 \leqslant M \leqslant 101$, *before* storing an asterisk in Z_M.

8. Different wavelengths.   Modify the program to calculate and plot the four wave shapes: square, triangular, spiked, and profile, using wavelengths less than 2π. As noted on page 107, for an arbitrary wavelength λ, $2\pi kx/\lambda$ should replace kx as the argument in all sine and cosine terms appearing in the sums defining $F_N(x)$ (see equation 3.2). Note that the coefficients $a_k, b_k, k = 1, 2, \ldots, N$, do not need to be changed.

9. Wave forms having nonzero area.   In the present version of the program, all the wave forms have *zero* net area (equal areas above and below the x-axis). The reason for this is that the leading coefficient in the cosine sum, a_0, is presently assumed to be zero. Apart from this term, every other term in the sine and cosine sums in equation 3.1 has zero net area for the interval $-\pi \leqslant x \leqslant +\pi$. To generate Fourier synthesized wave forms having some particular nonzero area, we obtain the value of a_0 from

$$a_0 = \frac{1}{2\pi} \int_{-\pi}^{+\pi} F(x)\, dx.$$

Note that the a_0 coefficient in equation 3.1 multiplies the function $\cos 0x = 1$, so that we may write equation 3.1 as

$$F(x) = a_0 + \sum_{k=1}^{\infty} a_k \cos kx + \sum_{k=1}^{\infty} b_k \sin kx.$$

Modify the program to Fourier synthesize several wave forms having some particular nonzero area and discuss the results.

10. Calculation of the Fourier coefficients.   Write a program to calculate the Fourier coefficients $a_1, a_2, \ldots, a_N; b_1, b_2, \ldots, b_N$, using equations 3.3 for any function $F(x)$ that can be read from data cards as a table of numbers. Run the program using some function originally defined in terms of a graph, from which you can read values of $x_1, x_2, \ldots, x_M; F(x_1), F(x_2), \ldots, F(x_M)$. Using the

---

* There are cases where the series converges even if this condition is not satisfied; for example, if the coefficients have alternating signs or if

$$a_1 = a_2 = a_3 = \ldots = a_N = \frac{1}{N} \quad \text{(the spiked wave)}.$$

Fourier coefficients calculated by your program, put these values into the
Fourier synthesis program and run the program using various numbers of
waves N.

Wave Pattern Program

11. Doppler Effect.   Using the wave pattern program, generate a wave pattern for a single source moving at a speed v, for which $v/c < 1$. Use the pattern
you obtain to check the validity of the Doppler Effect formula (equation 3.18)
for the case of a source moving directly toward or away from an observer. In
addition, use the pattern to determine the approximate variation of observed
frequency as a function of position, as a source goes past an observer, on a line
some distance away. (Use the technique outlined on page 128).

12. Shock waves.   Generate a shock wave pattern for a source moving at
some velocity v for which $v/c > 1$. With the aid of the pattern, check the shock
wave formula (equation 3.19).

13. Simple interference patterns.   Generate several simple interference patterns using sources of equal strengths and equal wavelengths. Make one of the
patterns with a source separation which is an appreciable fraction of the size
of the page (say $d = 10$), and another with a much smaller source separation
(say $d = 2$). Discuss the significant features of the patterns, including the maximum and minimum intensity regions, and compare your observations with
theory. You should choose the values of the parameters wisely, so that the
interesting features of the plots will be readily apparent, given the limitations
of the printer as a plotting device.

14. Complex interference patterns.   Generate several complex interference
patterns for which the sources have unequal strength and/or wavelengths. Observe and try to explain significant features in the patterns.

15. Incoherent sources.   By definition, the resultant intensity can be found
using $I = (A_1 + A_2)^2 = A_1^2 + A_2^2 + 2A_1A_2$. In the case of two incoherent sources,
the relative phase between the sources is constantly changing, so that the long
term time-average of the cross term $2A_1A_2$ is zero. Thus, to find the resultant
intensity, we may simply add the individual intensities $I = A_1^2 + A_2^2 = I_1 + I_2$.
Modify the program to produce wave patterns for incoherent sources, and compare the results with the patterns for coherent sources corresponding to the
same parameters.

16. More than two sources.   Modify the program so that it will produce interference patterns for a row of N equally spaced equal strength sources. Try
running the modified program using several values for the parameters and
compare your results with theory.

17. Shock waves and sonic boom.   Modify the program so that it prints out
numerical values for the intensity at a series of points on a line above or below
the x-axis. Run the program for several values of the velocity of the source to
study how the intensity of the shock wave varies with both position and velocity.
For this purpose, there is no need to generate the actual wave pattern, since you
only need numerical values for the intensity.

18. Interference with plane waves.   The amplitude of a plane wave at an arbitrary point in the xy-plane can be written

$$A(x, y) = S \sin \frac{2\pi}{\lambda} (x \cos \phi + y \sin \phi).$$

As before, S and λ are the source strength and wavelength, respectively. The angle ϕ is the direction of propagation of the waves with respect to the x-axis.

Modify the program so that it generates interference patterns for the case of a plane wave interfering with waves from a point source. Another interesting possibility is interference between two plane waves traveling along different directions. Run the modified version of the program using several choices for the parameters and discuss the results.

19. Two-dimensional Fourier analysis.   Modify the program so that it gives the intensity at each point I(x, y) according to

$$I(x, y) = |F_N(x, y)|^2,$$

where $F_N(x, y)$ is the two-dimensional Fourier synthesized wave form involving N waves. In order to generate an aesthetically pleasing picture, we choose the wavelength of the first harmonic to correspond to the size of the x or y interval represented in the plot: $-10 \leqslant x \leqslant +10$, i.e., $\lambda = 20$. In this case, the function $F_N(x, y)$ may be written

$$F_N(x, y) = \sum_{k=1}^{N} \sum_{l=1}^{N} a_{kl} \cos \frac{2\pi}{20} (kx + ly) + \sum_{k=1}^{N} \sum_{l=1}^{N} b_{kl} \sin \frac{2\pi}{20} (kx + ly).$$

One interesting choice for the coefficients is

$a_{kl} = 1/N$ for $k = 1, 2, 3, \ldots, N,\ 1 = 0,$

$a_{kl} = 0$ otherwise,

$b_{kl} = 0$ for all k and 1.

These correspond to the Fourier coefficients used in the one-dimensional spiked wave form. Another interesting choice is

$a_{kl} = 1/N$ for $k = 1, 2, 3, \ldots, N,\ 1 = 0;$

                      and for $k = 0,\ 1 = 1, 2, 3, \ldots, N,$

$a_{kl} = 0$ otherwise,

$b_{kl} = 0$ for all k and 1.

These coefficients will produce a "two-dimensional" spike.

Program to Solve the Schrödinger Equation

20. Single square well potential.   Using the technique discussed previously, run the program to find all the eigenvalues and eigenfunctions for a number of square well potentials having the same width W and a range of depths V. Compare the energy levels for each well, taking all energies with respect to the bottom of the well (that is, use $E - V_0$). Also compare your results with the energy levels for a well having infinitely high sides, for which the nth level is a distance above the bottom given by

$$\lim_{V_0 \to -\infty} (E_n - V_0) = \frac{n^2 \pi^2}{w^2} .$$

This result can be obtained using the wavelength of the wave function inside the well, defined as

$$\lambda = \frac{2\pi}{\gamma_1} = \frac{2\pi}{\sqrt{E - V_0}} ,$$

together with the boundary conditions for an infinite well which require that exactly an integral number of half-wavelengths fit inside the well, in order that the wave function have a node at each side:

$$n \frac{\lambda}{2} = w$$

21. Multiple square well potential. Using the technique discussed on pages 155-158 run the program to find all the eigenvalues and eigenfunctions for a number of double well potentials having different well separations S. All the double wells should have the same depth and width as one of the single well potentials used for problem 20, which has two or more eigenvalues. Discuss the relationship you observe between the double well eigenvalues and the eigenfunctions, and those of the single well.

Find all the eigenvalues and eigenfunctions for a potential consisting of a large number of wells, say NW = 10. Compare the results with those of the single well having the same well parameters.

22. Minimum well depth required for a bound state. Run the program to empirically determine the smallest well depth for which a single eigenfunction is obtained given a fixed well width. The procedure here is to keep the well width and trial energies fixed at some values, and vary the well depth V, until a first eigenfunction is obtained. Suitable values are W = 10., EMAX = 0., EMIN = $-.01$, NE = 2. Since the trial energies are very small negative numbers, the value of V for which E is a first eigenvalue is the well depth resulting in a state which is "barely bound." (A well of slightly smaller depth would have no bound state eigenvalues.)

Repeat the procedure for wells having other widths, and discuss how the minimum well depth required for a bound state is related to the width of the well.

23. Wells having unequal widths and depths. Modify the program so that the well parameters for each of a number of wells can be specified separately. Run the modified version of the program for the case of a double well, using several (unequal) values for the width and depth of each well. Discuss the results you obtain for the eigenfunctions and eigenvalues.

24. Eigenvalues for the semi-infinite square well potential. The "semi-infinite" square well potential is depicted in Figure 3.46, together with its first two eigenfunctions ψ_1 and ψ_2. The boundary condition $\psi(0) = 0$ is required by the fact that the potential rises to infinity at x = 0.* The wave function $\psi(x)$ inside the well consistent with this boundary condition is

$$\psi_{in}(x) = A \sin \gamma_1 x.$$

Outside the well, in order that $\psi(x)$ remain finite as x approaches infinity, only the decaying exponential must be present, so that we have the eigenfunction

$$\psi_{out}(x) = B e^{-\gamma_2 x},$$

where $\gamma_1 = \sqrt{V_0 - E}$ and $\gamma_2 = \sqrt{-E}$. Continuity of the eigenfunction and its first derivative at the well boundary (x = w) requires that

$$A \sin \gamma_1 w = B e^{-\gamma_1 w}$$

$$\gamma_1 A \cos \gamma_1 w = -\gamma_2 B e^{-\gamma_2 w}.$$

Figure 3.46   First two eigenfunctions for a semi-infinite square well

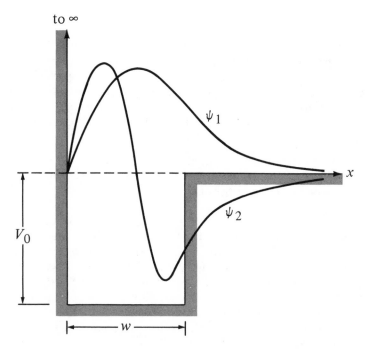

---

* Note that due to the boundary condition $\psi(0) = 0$, the eigenvalues for the semi-infinite square well of width w and depth V_0, are identical with the even eigenvalues $E_2, E_4, E_6, \ldots,$ for the ordinary square well of width 2w and depth V_0. The reason is that the even eigenfunctions are antisymmetric and therefore also satisfy $\psi(0) = 0$; moreover, the form of both potentials is identical for x greater than zero. You can therefore check the validity of your results by comparing them with those obtained for the ordinary square well.

Taking the ratio of these two equations, we obtain

$$\tan \gamma_1 w = -\frac{\gamma_1}{\gamma_2},$$

which we may write as

$$\tan \left[(V_0 - E)^{1/2} w \right] = -\left(\frac{E - V_0}{E} \right)^{1/2}.$$

The values of E which satisfy this equation are the eigenvalues of the semi-infinite square well. To solve the equation, we need to find the zeros in the function $F(E)$:

$$F(E) = \tan \left[(V_0 - E)^{1/2} w \right] + \left(\frac{E - V_0}{E} \right)^{1/2}.$$

Write a program that uses an iterative method, such as the decision tree algorithm discussed in Chapter 1, to find all the eigenvalues for the semi-infinite square well for several choices of the well parameters V_0 and w.

25. Solution of the Schrödinger equation for arbitrary potentials.   We can solve the Schrödinger equation for an arbitrary potential using a numerical technique, such as Euler's method or the improved Euler's method (see section 2.3). Let us first convert the second order Schrödinger equation

$$\frac{d^2 \psi}{dx^2} = (V(x) - E) \psi$$

into two first order equations:

$$\psi' = \frac{d\psi}{dx}$$

$$\frac{d\psi'}{dx} = (V(x) - E) \psi.$$

According to Euler's method, if we have initial values for ψ and ψ', then we can repeatedly find new values for ψ and ψ' from the previous ones using

$$\psi_{new} = \psi_{old} + \psi'_{old} \Delta x$$
$$\psi'_{new} = \psi'_{old} + (V(x) - E) \psi_{old} \Delta x.$$

Write a program to carry out this procedure, and find the eigenfunctions and eigenvalues for several potentials.

The technique for finding eigenvalues and eigenfunctions is the same as that used for the square well potential. Namely, an arbitrary choice is made for a trial energy, E, and we solve the Schrödinger equation using this trial energy. The solution will, in general, not be an eigenfunction, unless we happened to choose E to be one of the eigenvalues. We can determine whether E is above or below a particular eigenvalue depending on the number of nodes in the function. By choosing other trial energies and finding the number of nodes in each solution, we can then converge on particular eigenvalues.

Figure 3.47   *Three potential wells*

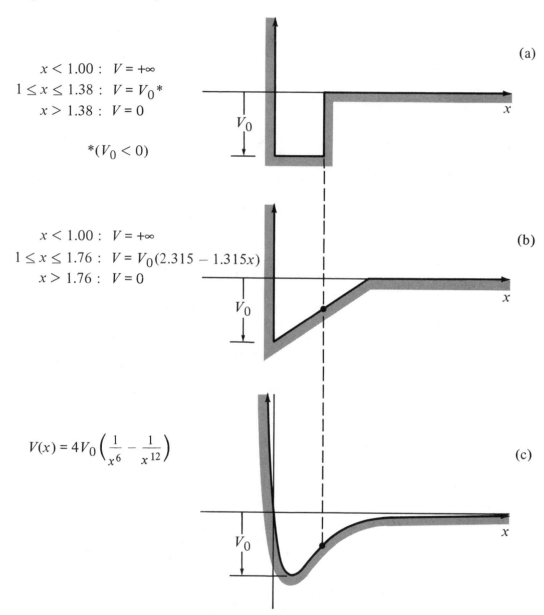

$x < 1.00 :\ V = +\infty$
$1 \le x \le 1.38 :\ V = V_0{}^{*}$
$x > 1.38 :\ V = 0$

$^{*}(V_0 < 0)$

(a)

$x < 1.00 :\ V = +\infty$
$1 \le x \le 1.76 :\ V = V_0(2.315 - 1.315x)$
$x > 1.76 :\ V = 0$

(b)

$V(x) = 4V_0 \left(\dfrac{1}{x^6} - \dfrac{1}{x^{12}} \right)$

(c)

Three potentials are shown in Figure 3.47. The first of these was previously discussed in problem 24. For this potential, the eigenfunctions can be expressed in analytic form using the sine and exponential functions. This permits you to check your numerical solutions and verify the correctness of the method. The other two potentials have been defined so that all three potentials have identical depths and widths. (The widths for the second and third potentials are defined in terms of the point at which the well reaches half its depth—see Figure 3.47

b, c.) The third one, known as the Lennard-Jones 6-12 potential, is important in molecular interactions. The Schrödinger equation has no analytic solution in this case and can only be solved numerically. To simplify the problem slightly, we assume that the steep slope of the Lennard-Jones potential for $x \leqslant 1$ is actually infinite. In this case, the boundary condition for all three potentials is $\psi(1) = 0$. An arbitrary choice of the initial slope $\psi'(1) = 1$ then determines the maximum height of the wave function.

4

Monte Carlo
Calculations

The Monte Carlo method is a problem solving technique which makes use of a sequence of random numbers. In physics, the most obvious applications are to processes such as radioactive decay and Brownian motion, in which randomness is an inherent part of the phenomenon. As we shall see, the Monte Carlo method is also useful in certain problems not related to random phenomena such as finding the center of mass of an object. The sequence of random numbers used in the Monte Carlo method could be generated by some random process, such as rolling dice, or drawing numbers from a hat. In practice, when we use the computer for a Monte Carlo calculation, we usually also have the computer generate the random numbers. We refer to such computer-generated numbers as *pseudorandom*, since no sequence of numbers generated according to a well-defined arithmetic procedure can be regarded as truly random.

4.1 Generation of Pseudorandom Numbers

In most methods for obtaining a sequence of pseudorandom numbers, each number in the sequence is used to find the next one using some arithmetic procedure.* Thus the entire sequence is determined given the first number. We note that any arithmetic procedure used to generate pseudorandom numbers on a computer yields a sequence that repeats itself after some number of terms. The reason is connected with the finite number of different numbers that can be represented on a computer owing to the finite size of each memory word. Let n designate the number of different numbers that can be represented. Then the maximum possible period of any pseudorandom number generator in which each number determines the next is equal to n, since any sequence of n + 1 numbers must contain at least one duplication and one duplication causes the entire sequence to repeat.

In selecting an arithmetic procedure for generating a sequence of pseudorandom numbers, we want the *period* of the sequence, i.e., the number of numbers before the sequence repeats itself, to be reasonably long compared to the maximum possible period determined by the computer word size. We also want the sequence to have all the statistical properties that a truly random sequence would have. A number of statistical tests will be discussed later.

The Power Residue Method

The power residue method is widely used and can generate a sequence of pseudo-

---

* In some methods, each number in the sequence is determined using more than one of the preceding numbers.

random integers, which has both a long period and good statistical properties. Given a first integer x_0, each number in the sequence can be found from the preceding one, according to

$$x_n = cx_{n-1}(\bmod N), \tag{4.1}$$

where c and N are two constant integers. The notation y = z (mod N) means that if z exceeds N, then the *modulus* N is subtracted from z as many times as necessary in order that $0 < y < N$; that is, y and z differ by some multiple of N (for example, 5752 (mod 100) = 52).

As an example, let us take N = 32, c = 5, and x_0 = 1. We can then repeatedly use equation 4.1 to find the entire sequence:

$$x_1 = 5x_0 \ (\bmod \ 32) = (5)(1) - (0)(32) = 5$$
$$x_2 = 5x_1 \ (\bmod \ 32) = (5)(5) - (0)(32) = 25$$
$$x_3 = 5x_2 \ (\bmod \ 32) = (5)(25) - (3)(32) = 29$$
$$x_4 = 5x_3 \ (\bmod \ 32) = (5)(29) - (4)(32) = 17$$
$$\vdots$$

The first 12 numbers in this sequence are

$$5, \quad 25, \quad 29, \quad 17, \quad 21, \quad 9, \quad 13, \quad 1, \quad 5, \quad 25, \quad 29, \quad 17.$$

We notice that the sequence repeats after eight numbers, which would be much too small a period for the sequence to be of any practical use. In general, the length of the period depends on all three parameters N, c, and x_0, so they must be chosen carefully to insure a long period. The length of the period increases with N, and we therefore wish to choose N as large as possible. The form of equation 4.1 insures that none of the numbers in the sequence exceeds N − 1. On a binary computer of b bits, the largest integer that can be represented is $2^{b-1} - 1$,* so that the largest modulus we can use is 2^{b-1}. In the example previously considered, the choice N = 32 would be suitable for a six bit computer which could represent any integer between −32 and +31.**

Let us now see how the choice of the other two parameters c and x_0 affect the length of the period. In the example, we used N = 32, c = 5, x_0 = 1, and obtained a sequence having a period of eight numbers. It can be shown that any other choice of x_0 (for the same c and N) would yield the same period if x_0 is odd, and half this period if x_0 is even. For example, with x_0 = 2, we obtain the sequence

$$2, \quad 10, \quad 18, \quad 26, \quad 2, \quad 10, \quad 18, \quad 26, \quad 2, \quad 10, \dots,$$

which has a period of four numbers. The length of the period is also affected by our choice of the constant c (here assumed to be 5). It can be shown that the longest period is obtained when c has the form c = 8n ± 3, where n is any positive integer. It can also be shown that the longest period attainable with the power residue method is N/4, if N is a power of 2. Thus, in the previous example,

---

* The reason the exponent is b − 1 instead of b is that all computers use one bit for the sign of an integer, leaving b − 1 for the number itself.

** Any computer having a word size of only six bits would probably represent each integer using two or more words. In this case, N could be increased to one more than the largest integer that can be represented on the computer, e.g., 2^{12-1}, if two words are used for each integer.

the choice $N = 32$, $c = 5$, $x_0 = 1$, for which c and x_0 satisfy the previously noted conditions, does give the longest period for this N. We summarize the conditions that N, c, and x_0 must satisfy to obtain the longest period:

N = one more than the largest integer that can be represented on the computer (On a binary computer of b bits, for which each integer is stored in a single word, we have $N = 2^{b-1}$.)

c = $8n \pm 3$, where n is any positive integer

x_0 = any odd integer.

(In order that the sequence have the best statistical properties, it is further recommended that the value of c be close to $N^{1/2}$, although the period is not thereby lengthened.)

In many applications involving random numbers, we do not want a sequence of pseudorandom integers; instead we want a sequence of numbers all of which lie in the interval from zero to one. This can be achieved by dividing each integer in the sequence by $N - 1$, which is the largest possible value.

Statistical Tests of Randomness

Of the two criteria for a good random number generator, it is much easier to satisfy the first, i.e., the sequence should have a long period, than the second, i.e., the sequence should have all the statistical properties of a truly random sequence. There is, unfortunately, no theoretical way to predict all the statistical properties of most pseudorandom sequences. Hence, we must rely on a number of statistical tests which can be empirically applied to particular pseudorandom sequences. The power residue method has yielded good results for the three tests described below.

1. Frequency Distribution Test.   The numbers should be consistent with a uniform (flat) frequency distribution from zero to one (assuming this to be their range). The test consists of counting the number of numbers which lie in each of M equal size subintervals between zero and one. If x_j is the number in the jth subinterval and $\bar{x}$ is the average number in all subintervals, then the *deviations* $d_j = x_j - \bar{x}$ should be consistent with the laws of statistics.

As an illustration, the first 300 numbers generated using the power residue method with $N = 2^{16}$, $c = 899$, and $x_0 = 13$ are listed in Figure 4.1. The program used to generate these numbers will be discussed later. The frequency distribution, or *histogram,* in Figure 4.2 shows the number of numbers in each of the ten subintervals between zero and one. The figure also indicates the deviations from a uniform distribution (exactly 30 in each subinterval). The laws of statistics applied to a truly random sample predict that as long as the average number of numbers in each subinterval is greater than about 25, the probability of finding x numbers in a subinterval is given by the Gaussian distribution $e^{-(x-\bar{x})^2/2\sigma^2}$, where σ is the *standard deviation* and is here given by $\sigma = \sqrt{\bar{x}}$. According to this distribution, the most likely number of numbers in any subinterval is the average number $\bar{x} = \sum\limits^{M} x_j / M$, and there is a 68% probability of finding x within a one standard deviation range of the average value: $\bar{x} - \sigma$ to $\bar{x} + \sigma$ (which corresponds to deviations in the range $-\sigma$ to $+\sigma$). For an average number $x = \sum\limits^{10} x_j / 10 = 30$, we find a standard deviation $\sigma = \sqrt{30} \cong 5.5$. As can be seen in Figure 4.2, seven out of ten of the deviations are less than 5.5, which is in good agreement with the expected 68%.

Figure 4.1

NUM= 300.00   N= 32768.00   X= 13.00   C= 899.00

.35667	.63640	.10837	.57726	.93878	.06265
.32469	.11582	.11698	.84051	.40434	.50969
.19700	.90515	.29637	.42424	.62188	.95087
.80175	.75286	.80151	.53337	.51323	.62053
.15799	.03256	.72655	.84777	.12528	.62194
.89599	.52696	.27464	.89032	.37010	.70794
.41673	.62908	.47612	.01816	.32402	.71941
.73241	.41984	.57256	.71374	.37059	.85314
.95392	.45476	.81347	.71203	.09305	.34977
.43602	.03183	.61504	.90368	.38676	.31242
.85522	.81951	.72021	.55431	.30741	.35582
.13181	.49309	.72784	.30454	.22312	.57952
.96899	.90173	.63091	.83001	.15793	.02231
.05521	.36955	.21409	.45866	.32517	.32316
.51244	.66619	.88525	.18418	.42796	.72527
.00003	.02744	.66436	.76092	.04440	.08164
.61058	.89813	.60649	.77825	.37236	.73815
.42235	.32286	.23807	.02280	.49419	.25987
.38841	.83093	.98102	.09220	.88208	.96258
.66314	.14170	.61791	.51744	.83447	.16349
.97095	.85418	.11332	.13272	.68389	.79772
.86865	.89105	.97150	.65203	.15525	.43663
.51689	.67174	.12156	.72521	.05490	.64391
.85748	.84973	.11887	.86053	.40678	.31486
.95001	.03354	.15140	.10654	.22336	.79901
.28367	.98889	.98614	.51695	.72662	.79290
.79662	.14371	.80718	.36369	.94653	.90594
.41697	.84857	.83862	.10532	.67925	.62755
.15213	.76501	.27927	.93957	.65069	.94812
.33256	.96252	.71801	.52965	.13968	.42863
.67119	.61541	.76714	.35881	.55687	.61199
.16013	.04697	.22294	.41490	.98297	.33634
.36448	.34019	.17869	.63372	.30595	.96112
.01999	.02988	.85919	.38609	.91601	.52940
.07981	.25669	.75835	.26017	.11405	.52574
.62798	.53624	.06571	.93182	.31791	.20646
.40019	.24088	.54686	.38682	.25755	.47349
.34129	.80902	.71758	.91375	.55962	.91888
.04953	.52751	.21921	.93225	.06619	.49284
.94733	.38078	.31028	.93475	.31590	.01718
.55394	.02182	.38377	.99896	.03977	.25181
.36869	.55412	.14280	.36979	.43358	.22666
.23801	.03207	.83453	.21836	.29960	.66765
.20212	.29600	.09500	.59392	.08158	.66546
.22678	.12827	.68944	.20908	.04074	.62615
.89013	.20548	.72192	.98212	.89550	.96594
.35478	.93542	.91949	.59825	.18577	.99872
.17972	.43352	.28153	.90936	.51030	.74572
.38163	.92157	.46385	.98920	.73956	.83972
.88238	.76312	.01975	.24937	.17386	.29893

Another test is to compute the *root mean square* (rms) *deviation* for all sub-intervals

$$d_{rms} = \sqrt{\frac{1}{M} \sum^{M} d_j{}^2} = \sqrt{\frac{1}{M} \sum^{M} (x_j - \bar{x})^2}.$$

For a large number of subintervals we expect d_{rms} to approach the standard deviation σ. In the present example we determine d_{rms} using $M = 10$ and $d_1 = 0$, $d_2 = 0, \ldots, d_{10} = +7$, as indicated in Figure 4.2. The computed value $d_{rms} = 4.3$ is reasonably close to the standard deviation $\sigma = 5.5$. (Reasonably close is here defined in terms of the expected "deviation of the deviation" given by $\pm\sigma/\sqrt{M} = \pm1$).

Figure 4.2

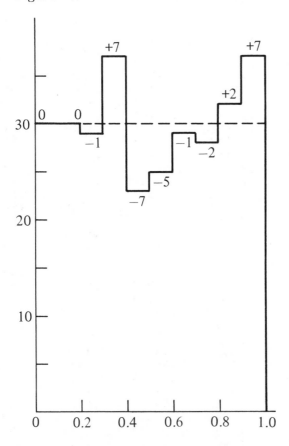

Other methods of testing that a histogram is consistent with an expected frequency distribution, such as the chi square test discussed in Chapter 5, can also be used to see if the set of pseudorandom numbers is distributed uniformly. However, a frequency distribution test alone is far from a proof of randomness of the sequence of numbers. One problem concerns the imprecision inherent in any statistical test, e.g., the rms deviation is only expected to be approximately equal to the standard deviation. Another more serious limitation of the frequency distribution test is its insensitivity to regularities in the *order* of numbers in the sequence. For example, if the 300 numbers listed in Figure 4.1 had been generated in a monotonically increasing sequence, we certainly would not regard this as a random sequence, but we would have obtained the identical histogram.

2. Frequency of Runs Test. Some information on the order of numbers in a sequence can be found by looking for "runs" of various length. A *run* of length k consists of a string of k consecutive numbers which is either monotonically increasing or decreasing, e.g.,

$$\ldots r_j > r_{j+1} < r_{j+2} < r_{j+3} < \ldots < r_{j+k} > r_{j+k+1} \ldots$$

and

$$\ldots r_j < r_{j+1} > r_{j+2} > r_{j+3} > \ldots > r_{j+k} < r_{j+k+1} \ldots$$

are both runs of length k. The test involves counting the occurrences of runs of different lengths appearing in a given sequence, and comparing with expected results. The expected results, based on a truly random sample, are $\frac{1}{12}(5N + 1)$ runs of length k = 1, $\frac{1}{60}(11N - 14)$ runs of length k = 2, and

$$\frac{2}{(k + 3)!} \left[N(k^2 + 3k + 1) - (k^3 + 3k^2 - k - 4) \right]$$

runs of length k, for k < N − 1. The expected number of runs of any length is $\frac{1}{3}(2N - 1)$.

3. "Above and Below the Mean" Run Test. Another type of run test is to look for strings of numbers which are either all above $\frac{1}{2}$ or all below $\frac{1}{2}$. Again, we can count the occurrences of runs of various lengths and compare with expected results. The expected number of runs of length k, for a truly random sample, is $(N - k + 3) 2^{-k-1}$, and the expected number of runs of any length is $\frac{1}{2}(N + 1)$.

Should a particular sequence of pseudorandom numbers give satisfactory results for the three tests we have just described, we would have some confidence in the method used to generate the sequence. However, it would certainly not prove that the sequence has all the statistical properties of a truly random sequence. There are, unfortunately, an unlimited number of possible regularities that a sequence of numbers might have that would escape detection in these three tests. A sequence which appears to be random according to several tests may reveal its nonrandom character only in some special application. The difficulty in determining that a sequence of numbers has no regularities of any kind makes finding a good random number generator as much an art as a science. Some techniques produce sequences of pseudorandom numbers that are "more random" than others. Even for a particular technique, such as the power residue method, the randomness of the sequence may vary according to the choice of the constants N, c, and x_0.

Generation of Pseudorandom Numbers with a Nonuniform Distribution

It is often useful to be able to generate pseudorandom numbers which have some specific distribution other than the uniform (flat) distribution from zero to one. If the specified distribution is uniform, but for an interval other than [0, 1], then a linear function of the random numbers can be used. For example, to generate a *random variable** which is uniformly distributed between $-\pi$ and $+\pi$, we could use $z = 2\pi(r - \frac{1}{2})$, where r is a uniformly distributed random variable between zero and one.

In the more general case, we want to generate a random variable z whose distribution function $f(z)$ is nonuniform. We can accomplish this by first choosing a random variable r which is *uniformly* distributed in the interval $[0, 1]$. We can then find z from

$$F(z) \equiv \int_{-\infty}^{z} f(z')\,dz' = r,$$

where $F(z)$ is the *cumulative distribution function* corresponding to $f(z)$ (see Figure 4.3). Since r is uniformly distributed between zero and one, the probability that $F(z)$ lies in any interval Δy is given by $\Delta p = \Delta y$. As can be seen from the plot of $F(z)$ against z (Figure 4.3b), this is also the probability that z lies in a corresponding interval Δz, where

$$\Delta z = \Delta y \left/ \frac{dF}{dz} \right. = \Delta y / f(z),$$

so that

$$\frac{\Delta p}{\Delta z} = \frac{\Delta p}{\Delta y} \cdot f(z) = f(z),$$

which shows that z has the desired distribution function $f(z)$. (Note how the uniformly distributed marks on the vertical axis in Figure 4.3b correspond to marks on the horizontal axis having the distribution function $f(z)$.)

Figure 4.3   A distribution function f(z) and its cumulative distribution function F(z)

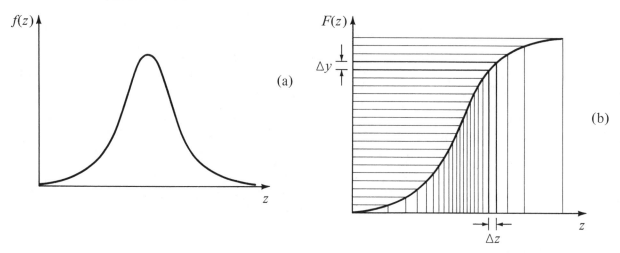

---

* A variable z is called a random variable if the probability of it assuming particular values is given by some function $f(z)$ known as its *distribution function*.

For example, suppose we wish to generate a random number z_0 which is distributed according to the distribution function $f(z_0)$, where

$$f(z_0) = e^{-az_0} \quad \text{for } 0 \leqslant z_0 \leqslant +\infty$$

$$f(z_0) = 0 \qquad \text{for } z_0 < 0.$$

The cumulative distribution function is

$$F(z_0) = \int_0^{z_0} f(z)\,dz = \frac{1}{a}(1 - e^{-az_0}).$$

We now must solve the equation $F(z_0) = r$ to obtain the inverse function $z_0 = F^{-1}(r)$. Using

$$F(z_0) = \frac{1}{a}(1 - e^{-az_0}) = r,$$

we find with the aid of a little algebra that

$$z_0 = -\frac{1}{a} \ln(1 - ar) = F^{-1}(r).$$

The variable z_0 is distributed according to the distribution function $f(z_0) = e^{-az_0}$, provided r is distributed according to the uniform distribution from zero to one.

Program to Generate Pseudorandom Numbers

The program given in flow diagram form in Figure 4.4 and listed on page 187, uses the power residue method to generate a specified number of pseudorandom numbers having a uniform distribution between zero and one. The program first reads a data card containing numerical values for the parameters NUM, N, X, and C, where

NUM = Number of pseudorandom numbers to be generated
N = the modulus
X = x_0, the first pseudorandom number in the sequence
C = c, the constant appearing in the power residue formula.

The program generates the random numbers in sets of six, in order that each set of six numbers can be printed on a single line. Thus, instead of a single loop from 1 to NUM, the program uses two nested loops, the outer one going from 1 to NUM/6, and the inner one going from 1 to 6. To obtain each pseudorandom number from the preceding one, we need to compute the product, cx (modulo N). This can be done by first computing the new x from the preceding one using $x \leftarrow cx$. Then, to find x (modulo N), we can use

$$x \leftarrow x - N(x/N), \tag{4.2}$$

where all operations are carried out according to the rules of integer arithmetic, i.e., quantities are truncated to the next lower integer after each arithmetic operation. A complication may arise when the product cx is first computed: an

Figure 4.4   Flow diagram to generate a sequence of pseudorandom numbers

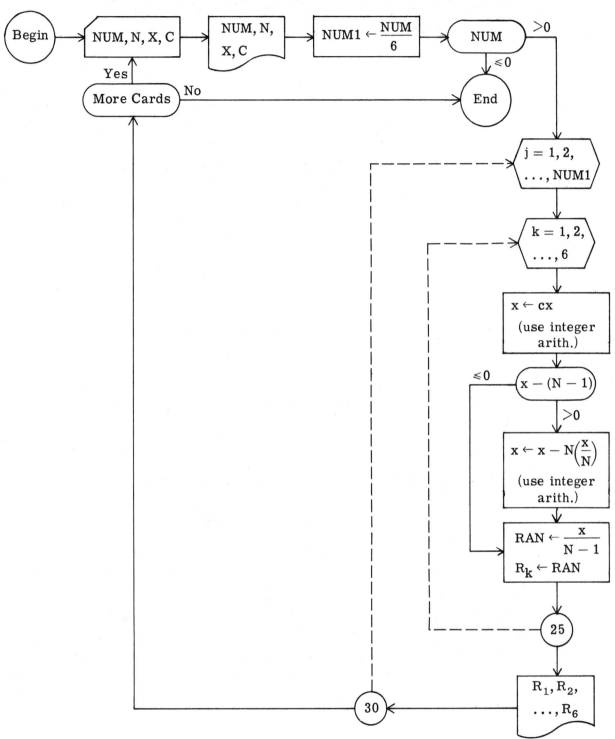

```
C                 PSEUDORANDOM NUMBER GENERATOR
C
C
C----------------------------------------------------------------------
C
C                 THIS PROGRAM GENERATES PSEUDORANDOM NUMBERS USING THE POWER
C                 RESIDUE METHOD, ACCORDING TO WHICH EACH NUMBER IS OBTAINED
C                 FROM THE PRECEDING ONE USING THE FORMULA X = C*X (MOD N).
C
      REAL NUM,N,N1
      DIMENSION R(6)
C
C                 NUM=NUMBER OF PSEUDORANDOM NUMBERS DESIRED
C                 N=MODULUS
C                 X=FIRST NUMBER IN THE SERIES
C                 C=THE CONSTANT IN THE POWER RESIDUE FORMULA
C
    1 READ(2,100)NUM,N,X,C
      WRITE(3,101)NUM,N,X,C
      N1=N-1.0
      M=N
      IX=X
      IC=C
      NUM1=NUM
      NUM1=NUM1/6
      IF(NUM)50,50,2
C
C                 GENERATE NUM RANDOM NUMBERS, 6 AT A TIME
C
    2 DO 30 J=1,NUM1
      DO 25 K=1,6
C
C                 THE NEXT STATEMENT GIVES IC*IX  (MOD N)  ON COMPUTERS FOR
C                 WHICH NUMBERS BIGGER THAN N ARE NOT PERMITTED
C
      IX=IABS(IC *IX)
      IF(IX-N1)20,20,10
C
C                 THE NEXT STATEMENT GIVES IC*IX  (MOD N) ON COMPUTERS FOR
C                 WHICH NUMBERS BIGGER THAN N ARE PERMITTED
C
   10 IX=IX-M*(IX/M)
   20 RAN=IX
C                 DIVIDE BY N-1, SO THAT RAN LIES BETWEEN ZERO AND ONE
      RAN=RAN/N1
   25 R(K)=RAN
   30 WRITE(3,102)(R(K),K=1,6)
      GO TO 1
   50 CALL EXIT
  100 FORMAT(4F10.5)
  101  FORMAT(5H1NUM=,F10.2,6H     N=,F10.2,6H     X=,F10.2,4H  C=,F10.2/)
  102 FORMAT(6F10.5)
      END
```

"overflow" may occur, i.e., the product may exceed the largest integer that can be represented on the computer, I_{max}. Due to the way most computers carry out integer arithmetic in FORTRAN, if there is an overflow, the leading (most significant) part of the result is dropped.* In this case, the result is computed modulo $(I_{max} + 1)$, apart from its sign which may come out negative, if an overflow into the leading (sign) bit occurs. Thus, by using $x \leftarrow |cx|$, we get a result which is *automatically* computed modulo $(I_{max} + 1)$. This implies that the subsequent computation of x (modulo N) using equation 4.2 *can only be correct if* $I_{max} + 1$ *is divisible by* N. On a binary computer this restricts N to powers of two. In other words, if $N = 2^b$, then the possible initial loss of leading bits (due to overflow) in no way affects the computation of x (modulo N), in which all but the lower order b bits are dropped.

Finally, to obtain a pseudorandom number between zero and one, we divide x by $N - 1$, its maximum possible value. After the program finds six consecutive numbers, $R_1, R_2, \ldots, R_6$, it prints them on a line and proceeds to the next set of six. When the entire sequence of NUM numbers has been printed, the program reads the next data card if any remain.

The sample results shown in Figure 4.1 have been already discussed. This output can be generated using a data card containing the following values for the parameters:

NUM	N	X	C
300.	32768.	13.	899.

Problems 1-4 at the end of this chapter pertain to the generation of pseudorandom numbers.

4.2   Monte Carlo Evaluation of Definite Integrals

One application of the Monte Carlo method involves the use of random numbers to evaluate a definite integral. To illustrate the technique, Figure 4.5a shows a plot of some function f(x) that we wish to integrate between limits $x = a_1$ and $x = a_2$. The value of the integral

$$\int_{a_1}^{a_2} f(x)\,dx$$

is equal to the shaded area under the curve between $x = a_1$ and $x = a_2$. We can obtain an approximate value for this area using a large number of points at random locations within the rectangle shown in Figure 4.5b. If we know A, the area of the rectangle, and if we count the number of points N inside the rectangle and the number of points n that lie under the curve, then we can obtain an approximate value for a, the area under the curve:

$$a = fA,$$

where $f = n/N$ is the fraction of all points that lie under the curve. Thus, to

---

*   This is not universally true although it is the way an overflow is treated on most computers.

evaluate a definite integral, we need only determine what fraction of a large number of randomly located points lies under the curve of the function to be integrated. We extend this idea to multiple integrals in the following sections.

Figure 4.5   Evaluating an integral using the Monte Carlo method

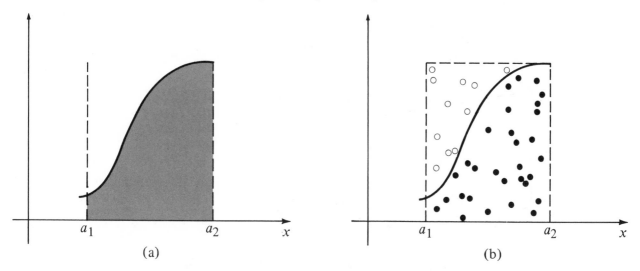

(a)   (b)

Calculation of the Mass and Center of Mass Coordinates of an Object

As an illustration of the Monte Carlo technique to evaluate multiple integrals, we consider the problem of calculating the mass of an object of known shape. We assume that the density ρ is a known function of position within the object. The mass of a small volume element $dV = dx\ dy\ dz$, is given by

$$dm = \rho\ dx\ dy\ dz.$$

The mass of the object can be calculated from the integral

$$M = \int dm,$$

which actually stands for the triple integral

$$M = \iiint \rho\ dx\ dy\ dz.$$

We can find the mass of the object and thereby evaluate this integral using the Monte Carlo technique. Let us imagine the object whose mass we wish to find encased in a rectangular block of known volume V. We consider the block a collection of a large number (N) of point masses occupying random locations within the block rather than a continuous substance. (See Figure 4.6.) (Our model corresponds to the atomic nature of the block for a sufficiently large N.) The simplest case to consider is a block of uniform density. In this case the mass of the rectangular block is given by

$$M = \rho V$$

and Δm, the mass of each of the N point masses making up the block, is given by

$$\Delta m = \frac{\rho V}{N} .$$ (4.3)

Figure 4.6   Randomly located point masses within a block

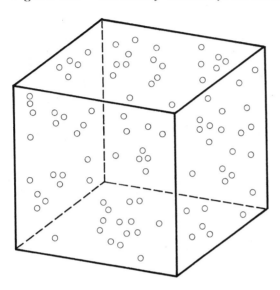

We can determine the mass of the object of known shape within the block by simply counting the number of randomly located point masses n that lie within the object (the solid points in Figure 4.7). The mass of the object is given by

$$M = n\Delta m,$$

which upon substitution of equation 4.3, becomes

$$M = \frac{n\rho V}{N} .$$

The same technique can be used even if the object is made of some material having a nonuniform density ρ which is a known function of position:

$$\rho = \rho(x, y, z).$$ (4.4)

In this case, according to equation 4.3, the jth point mass has a value

$$\Delta m_j = \frac{\rho_j V}{N} ,$$ (4.5)

where, for points inside the object, we determine the density at the jth point using equation 4.4, and for points outside the object, we use $\rho_j = 0$ (the empty circles in Figure 4.7). To find the total mass of the object we add the contributions from each point:

$$M = \sum_{j=1}^{N} \Delta m_j.$$ (4.6)

Figure 4.7   Determination of the mass of an object enclosed by the block

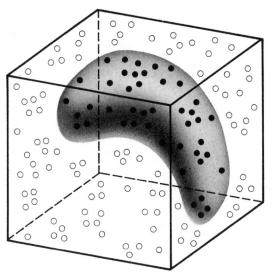

As a second illustration of how the Monte Carlo method can yield an approximate value for a multiple integral, we consider the problem of finding the center of mass coordinates of an object of known shape.

For a discrete set of point masses Δm_j located at positions (x_j, y_j, z_j), the coordinates of the center of mass are given by

$$x_{cm} = \frac{1}{M} \sum x_j \Delta m_j$$

$$y_{cm} = \frac{1}{M} \sum y_j \Delta m_j$$

$$z_{cm} = \frac{1}{M} \sum z_j \Delta m_j$$

which in vector form may be written

$$\mathbf{r}_{cm} = \frac{1}{M} \sum \mathbf{r}_j \Delta m_j .$$

For a continuous object this becomes

$$\mathbf{r}_{cm} = \frac{1}{M} \int \mathbf{r} \, dm \tag{4.7}$$

which actually stands for the triple integral

$$\mathbf{r}_{cm} = \frac{1}{M} \iiint \mathbf{r} \rho \, dx \, dy \, dz .$$

We can again approximate a continuous object using a collection of randomly located point masses in order to find an approximate value for the integral in equation 4.7 in a manner which is completely analogous to the way we can calculate an object's mass. While integrals such as those for the mass and center of mass coordinates of an object can also be evaluated by a more conventional numerical technique, such as the trapezoidal rule or Simpson's rule (see section 2.2), the Monte Carlo method may be easier to carry out. This is particularly true in problems involving multiple integrals where the region of integration (the shape of the object, in the present problem) makes it difficult to determine the limits of each integral. Both the Monte Carlo method and the conventional integration techniques yield only approximate answers. The accuracy depends on N, the number of randomly located points in the case of the Monte Carlo method, and the number of intervals in the case of the conventional methods. One disadvantage of the Monte Carlo method is its slow rate of convergence with increasing N. The average error in evaluating an integral using the Monte Carlo method is proportional to $N^{-1/2}$.

Program to Calculate the Mass and Center of Mass Coordinates of an Object

We assume that the object whose mass and center of mass coordinates we wish to determine is completely enclosed by a cube, one unit on a side, centered at the origin.* In order to generate points having random locations within the cube, the coordinates (x, y, z) of each point must be three random numbers between $-\frac{1}{2}$ and $+\frac{1}{2}$. Thus, we must subtract $\frac{1}{2}$ from each random number generated between 0 and 1 to obtain a coordinate.

For each point (x, y, z), we need to determine whether the point lies inside or outside an object of specified shape. In the present version of the program, the object is a sphere of radius $R = 0.5$, which has a vertical hole of radius $r = 0.3$ (see Figure 4.8). For an object of this shape, the point (x, y, z) lies inside the object only if both of the following conditions are satisfied:

$$R = (x^2 + y^2 + z^2)^{1/2} < 0.5$$
$$r = (x^2 + y^2)^{1/2} > 0.3.$$

We assume that the object has a uniform density of 1.0; if a randomly located point lies inside the object we use $\rho = 1.0$, and if it lies outside we use $\rho = 0$. Each of the N point masses has a mass given by equation 4.3. Since we have assumed that V, the volume of the cube which encloses the object is 1.0, we have

$$\Delta m = \frac{\rho}{N}.$$

The program is given in flow diagram form in Figure 4.9 and listed on pages 197-199. The program begins by reading from a data card numerical values for N and NRUNS, where,

N   = the number of randomly located points to use in calculating the mass and center of mass coordinates
NRUNS = the number of times to repeat the whole calculation using the same value of N.

---

*   In the general case, we would use a rectangular block whose dimensions are chosen to *just* enclose the object, in order to make the number of random points inside the rectangular block which also lie inside the object as large as possible, thereby minimizing the statistical uncertainty.

Figure 4.8   The object used in the present version of the program

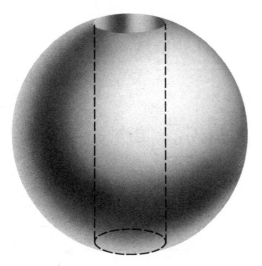

(Note that since different random numbers are used each time the calculation is repeated, the results will differ for each of the NRUNS runs. However, if the program is rerun another time, the results are reproducible.)

After reading the data card and initializing a number of parameters, the program enters a loop over the number of runs, NRUNS, which contains a loop over the number of point masses, N. For each point mass, it generates* three pseudo-random numbers R_1, R_2, and R_3, which are the coordinates x, y, and z. The program prints out these coordinates for the first 50 points to permit a check on their randomness. It then tests whether each point lies inside the specified object (a sphere with a hole in it, in the present case), and if so, it adds the contributions for this point into the sums by which the mass and center of mass coordinates are computed (equations 4.6 and 4.7). The program prints out the results and repeats the entire calculation (using different random numbers), as many times as have been specified by the value of NRUNS. After the specified number of runs are executed, it computes and prints out the average for all runs and the rms deviations of the mass and the center of mass coordinates r_{cm_k}, $k = 1, 2, 3$. It then reads the next data card, if any remain to be read.

Comments on Data Cards and Sample Results

The sample output shown in Figures 4.10-4.11 can be generated using three data cards containing the following values for the parameters:

---

*   The program uses the power residue method discussed in section 4.1 to generate a sequence of pseudorandom numbers uniformly distributed between zero and one. The constants N, c, and x_0, used in the power residue method, are assigned the values $N = 32768$, $c = 899$, $x_0 = 13$. If you wish to use some other random number generator, you need only replace those statements in the program contained between the dotted lines with other statements which generate a pseudorandom number between zero and one, and store it in the variable RAN.

Figure 4.9   *Flow diagram for Monte Carlo integration*

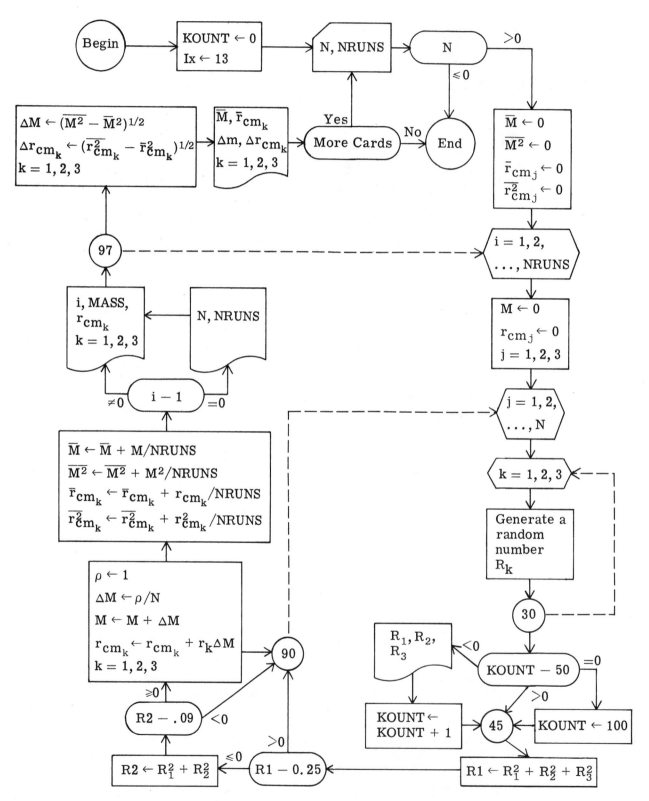

Figure 4.10

COORDINATES OF FIRST 50 RANDOM POINTS

X	Y	Z
-.14333	.13640	-.39163
.07726	.43878	-.43735
-.17531	-.38418	-.38302
.34051	-.09566	.00969
-.30300	.40515	-.20363
-.07576	.12188	.45087
.30175	.25286	.30151
.03337	.01323	.12053
-.34201	-.46744	.22655
.34777	-.37472	.12194
.39599	.02696	-.22536
.39032	-.12990	.20794
-.08327	.12908	-.02388
-.48184	-.17598	.21941
.23241	-.08016	.07256
.21374	-.12941	.35314
.45392	-.04524	.31347
.21203	-.40695	-.15023
-.06398	-.46817	.11504
.40368	-.11324	-.18758
.35522	.31951	.22021
.05431	-.19259	-.14418
-.36819	-.00691	.22784
-.19546	-.27688	.07952
.46899	.40173	.13091
.33001	-.34207	-.47769
-.44479	-.13045	-.28591
-.04134	-.17483	-.17684
.01244	.16619	.38525
-.31582	-.07204	.22527
-.49997	-.47256	.16436
.26092	-.45560	-.41836
.11058	.39813	.10649
.27825	-.12764	.23815
-.07765	-.17714	-.26193
-.47720	-.00581	-.24013
-.11159	.33093	.48102
-.40780	.38208	.46258
.16314	-.35830	.11791
.01744	.33447	-.33651
.47095	.35418	-.38668
-.36728	.18389	.29772
.36865	.39105	.47150
.15203	-.34475	-.06337
.01689	.17174	-.37844
.22521	-.44510	.14391
.35748	.34973	-.38113
.36053	-.09322	-.18514
.45001	-.46646	-.34860
-.39346	-.27664	.29901

	N	NRUNS
1st card	250.	2.
2nd card	250.	5.
3rd card	1000.	5.

We note that the x, y, z-coordinates of the center of mass are, as expected close to zero, due to the symmetrical shape of the object. The masses for each run are all reasonably close to the expected value 0.2780. (The mass of a sphere of radius a, which has a hole of radius c, is given by

$$M = \frac{4}{3} \pi (a^2 - c^2)^{3/2}\rho;$$
$$a = 0.5, c = 0.3, \rho = 1.0.)$$

The computed rms deviations* for each series of runs is a measure of the

Figure 4.11

N= 250.00 NRUNS=   2.00

| RUN | 1 MASS= | .25200 | C.M. COORDINATES | X= | -.00283 | Y= | -.02247 | Z= | .00133 |
| RUN | 2 MASS= | .26400 | C.M. COORDINATES | X= | .02495 | Y= | -.05291 | Z= | .02251 |

| AVERAGES | .25800 | | | .01106 | | -.03769 | | .01192 |
| DEVIATIONS | .00600 | | | .01389 | | .01522 | | .01059 |

N= 250.00 NRUNS=   5.00

RUN	1 MASS=	.29200	C.M. COORDINATES	X=	-.01835	Y=	.00900	Z=	.02982
RUN	2 MASS=	.24800	C.M. COORDINATES	X=	.07062	Y=	.02603	Z=	.01849
RUN	3 MASS=	.23600	C.M. COORDINATES	X=	.06336	Y=	.00349	Z=	-.00757
RUN	4 MASS=	.31600	C.M. COORDINATES	X=	.03069	Y=	-.03605	Z=	-.02286
RUN	5 MASS=	.27600	C.M. COORDINATES	X=	-.06796	Y=	-.04542	Z=	.01048

| AVERAGES | .27360 | | | .01567 | | -.00859 | | .00567 |
| DEVIATIONS | .02902 | | | .05229 | | .02744 | | .01876 |

N= 1000.00 NRUNS=   5.00

RUN	1 MASS=	.28600	C.M. COORDINATES	X=	-.01753	Y=	-.01019	Z=	-.00651
RUN	2 MASS=	.28700	C.M. COORDINATES	X=	.01192	Y=	-.00488	Z=	.00654
RUN	3 MASS=	.25800	C.M. COORDINATES	X=	-.00320	Y=	-.00659	Z=	.01955
RUN	4 MASS=	.25700	C.M. COORDINATES	X=	.00586	Y=	.01424	Z=	-.00109
RUN	5 MASS=	.24000	C.M. COORDINATES	X=	.00533	Y=	.03663	Z=	-.00171

| AVERAGES | .26560 | | | .00048 | | .00584 | | .00336 |
| DEVIATIONS | .01823 | | | .01021 | | .01758 | | .00911 |

----

* It can be shown with the aid of a little algebra that the following identity holds for the rms deviation:

$$d_{rms} = \sqrt{\frac{\sum\limits_{j}^{N} (x_j - \bar{x})^2}{N}} = \sqrt{\frac{\sum\limits_{j}^{N} x_j^2 - \left(\sum\limits_{j}^{N} x_j\right)^2}{N}},$$

where $\bar{x}$ is the average of the N values $x_1, x_2, \ldots, x_N$.

errors in each quantity. The differences between the computed values and the expected values of the mass and the center of mass coordinates are, in most cases, less than the rms deviations. (The laws of statistics predict this should be the case about 68% of the time—see page 180.) The laws of statistics also predict that the rms deviations should decrease as the number of random points N increases, according to the formula $N^{-1/2}$. Thus, we should find the standard deviations for $N = 1000$ are about half those for $N = 250$. This is roughly borne out by the results for the second and third data cards. The agreement would presumably be even better if a larger number of runs was used in each case.

Problems 5-11 at the end of this chapter pertain to the Monte Carlo evaluation of definite integrals.

```
C               MONTE CARLO EVALUATION OF INTEGRALS
C
C-----------------------------------------------------------------------
C
C               PROGRAM TO CALCULATE THE MASS AND CENTER OF MASS COORDINATES
C               OF AN OBJECT OF SPECIFIED SHAPE AND DENSITY USING RANDOMLY
C               LOCATED POINTS
C
        REAL N,NRUNS,MASS
        DIMENSION  R(3),RCM(3),DRCM(3),AVCM(3),AVCM2(3)
        KOUNT=0
C
C               AN ODD INTEGRAL VALUE FOR IX IS NEEDED TO INITIALIZE THE
C               RANDOM NUMBER GENERATOR.
C
        IX=13
        WRITE(3,1004)
C
C               READ THE VALUES OF N AND NRUNS FROM A DATA CARD.
C               N = NUMBER OF RANDOMLY LOCATED POINTS TO USE
C               NRUNS = NUMBER OF RUNS TO MAKE USING THIS VALUE OF N
C
   10   READ(2,1000)N,NRUNS
        IF(N)100,100,20
   20   MRUNS=NRUNS
        AVM=0.
        AVM2=0.
        DO 25 J=1,3
        AVCM(J)=0.
   25   AVCM2(J)=0.
C
C               LOOP OVER THE NUMBER OF RUNS.
C
        DO 97 I=1,MRUNS
        M=N
        MASS=0.
        RCM(1)=0.
        RCM(2)=0.
        RCM(3)=0.
C
C               LOOP OVER THE NUMBER OF POINTS.
C
        DO 90 J=1,M
C
C               LOOP OVER THREE COORDINATES PER POINT.
C
```

```
      DO 30 K=1,3
C
C                THE STATEMENTS BETWEEN THE DOTTED LINES GENERATE A RANDOM NUMBER.
C---------------------------------------------------------------------------------
C
      IX=IABS(899*IX)
      IF(IX-32767)5002,5002,5001
 5001    IX=IX-(32767+1)*(IX/(32767+1))
 5002 CONTINUE
      RAN=IX
      RAN=RAN/32767.
C
C---------------------------------------------------------------------------------
C
C                SUBTRACT 0.5 TO GET COORDINATES BETWEEN -0.5 AND +0.5.
C
  30  R(K)=RAN-0.5
C
C                IF LESS THAN 51 POINTS HAVE BEEN FOUND, WRITE OUT THE COORDINATES.
C
      IF(KOUNT-50)40,42,45
  40  WRITE(3,1003)R(1),R(2),R(3)
      KOUNT=KOUNT+1
      GO TO 45
  42  WRITE(3,1005)
      KOUNT=100
  45  CONTINUE
C
C                TEST TO SEE IF RANDOM POINT LIES INSIDE THE OBJECT.
C
      R1=R(1)**2+R(2)**2+R(3)**2
C
C                SEE IF POINT IS INSIDE SPHERE OF RADIUS 0.5.
C
      IF(R1-0.25)50,50,90
  50  R2=R(1)**2+R(2)**2
C
C                SEE IF POINT IS OUTSIDE CYLINDER OF RADIUS 0.3.
C
      IF(R2-0.09)90,55,55
C
C                IF BOTH CONDITIONS ARE SATISFIED, WE USE RHO = 1.0  AND ADD
C                THIS POINT'S CONTRIBUTION TO THE MASS AND TO THE CENTER OF
C                MASS COORDINATES (RCM(K)).
C
  55  RHO=1.0
      DM=RHO/N
      MASS=MASS+DM
      DO 80 K=1,3
  80  RCM(K)=RCM(K)+R(K)*DM
  90  CONTINUE
C
C                COMPUTE AVERAGES OF MASS, MASS**2, RCM, RCM**2 FOR ALL RUNS.
C
      AVM=AVM+MASS/NRUNS
      AVM2=AVM2+MASS**2/NRUNS
      DO 92 K=1,3
      RCM(K)=RCM(K)/MASS
      AVCM(K)=AVCM(K)+RCM(K)/NRUNS
```

```
 92    AVCM2(K)=AVCM2(K)+RCM(K)**2/NRUNS
       IF(I-1)96,95,96
 95    WRITE(3,1001)N,NRUNS
 96    WRITE(3,1002)I,MASS,(RCM(K),K=1,3)
 97    CONTINUE
C
C                 COMPUTE RMS DEVIATIONS.
C
       DMASS=SQRT(AVM2-AVM**2)
       DO 98 K=1,3
 98    DRCM(K)=SQRT(AVCM2(K)-AVCM(K)**2)
       WRITE(3,1006)AVM,(AVCM(K),K=1,3),DMASS,(DRCM(K),K=1,3)
C
C                 GO READ NEXT DATA CARD.
C
       GO TO 10
1000   FORMAT(2F10.5)
1001   FORMAT(/96H =======================================================
      1=====================================================///3H N=,F10.2,7H NRUNS=,
      2 F10.2/)
1002   FORMAT(5H RUN ,I3,6H MASS=,F10.5,32H               C.M. COORDINATES
      1X=,F10.5,4H  Y=,F10.5,4H  Z=,F10.5)
1003   FORMAT(3F10.5)
1004   FORMAT(42H1COORDINATES OF FIRST 50 RANDOM POINTS     //
      128H       X             Y          Z/)
1005   FORMAT(1H1)
1006   FORMAT(/9H AVERAGES,5X,F10.5,28X,3F14.5/11H DEVIATIONS,3X,F10.5,
      1 28X,3F14.5)
 100   CONTINUE
       CALL EXIT
       END
```

4.3 Simulation of a Chain Reaction

A chain reaction, which can occur in certain radioactive substances such as uranium, is a random process which can be studied with the aid of the Monte Carlo method. The nucleus of an atom of the uranium isotope U^{235} is inherently unstable. By virtue of its instability, the nucleus spontaneously disintegrates or *fissions* into a number of fragments. The *half-life* of a radioactive substance is the amount of time required for half of a large number of nuclei to disintegrate; U^{235} has a half-life of 707 million years. The energy released per atom in the fission process is about a million times greater than the energy released during an ordinary chemical process, such as burning wood or coal. However, due to the long half-life, only a relatively small fraction of all nuclei in a piece of uranium undergo fission at any one time, so that the rate at which energy is released may only be sufficient to make the uranium slightly warm to the touch.

The rate at which energy is released is drastically accelerated in the event of a chain reaction, which can lead to a nuclear explosion. In a chain reaction, neutrons which are emitted during one spontaneous fission collide with other U^{235} nuclei. The other U^{235} nuclei absorb the neutrons, which causes them to become highly unstable and very rapidly undergo fission, thereby emitting more neutrons which trigger more fissions, and so on. We refer to each phase of this process as a *generation*. If we assume that two neutrons are emitted during each fission, and that every emitted neutron induces another fission, then starting with N spontaneous fissions, there will be 2N induced fissions after one generation, 4N after two generations, and 2^n N after n generations. Thus, the number of induced fissions grows exponentially, reaching $2^{30} \approx$ one billion times the original number of spontaneous fissions in only 30 generations.

In view of the preceding discussion, you may wonder why a chain reaction doesn't always result from spontaneous fissions. The answer is related to the notion of *critical mass,* which is the smallest mass of uranium or other fissionable material in which a self-sustaining chain reaction can occur. Due to the small size of the uranium nucleus, neutrons emitted during nuclear fission have to travel, on the average, appreciable distances (on the order of centimeters) before interacting with other nuclei and inducing them to fission. Thus, for a small piece of uranium, even though two neutrons may be emitted in each spontaneous fission, the average number of induced fissions f caused by the neutrons is some number less than two. Let us define the factor f more precisely. Suppose N nuclei undergo spontaneous fission thereby emitting 2N neutrons, of which N_{in} induce other nuclei in the piece of uranium to fission. (The remaining neutrons leave the piece of uranium before colliding with a nucleus.) During the first generation, the fractional change in the number of nuclei undergoing fission is given by

$$f = \frac{N_{in}}{N},$$

which we shall refer to as the *survival fraction*. After two generations, the number of induced fissions is given by $f^2 N$, and after n generations, the number is $f^n N$. Thus, the number of induced fissions grows exponentially if and only if f is greater than 1.0.

The value of f for a particular piece of uranium is determined by its mass, shape, and purity. A piece of uranium for which $f = 1.0$ is said to have a critical mass. If a piece of uranium has a mass greater than the critical mass M_c, it will spontaneously undergo a chain reaction and possibly cause a nuclear explosion. To create such an explosion, one need only bring together two pieces of uranium whose combined mass exceeds the critical mass. It is clearly very important to be able to determine the value of the critical mass theoretically, as the experimental determination is a bit risky.

Calculation of the Critical Mass

We can calculate the critical mass for a block of uranium "experimentally," by finding the survival fraction f for a range of masses and shapes. (Recall that a block has the critical mass provided $f = 1.0$.) The computer procedure we shall use to find the value of f for a block of specified size and shape involves generating a large number (N) of simulated random fissions, and keeping a count of the number (N_{in}) of induced fissions which are caused by the emitted neutrons. We illustrate the procedure by giving the steps necessary to generate one random fission, shown schematically in Figure 4.12.

We first choose the location of the nucleus undergoing fission to be a random point (x_0, y_0, z_0), lying within the boundaries of the piece of uranium. If we assume that the piece of uranium is a rectangular block of dimensions $a \times a \times b$,

Figure 4.12   Two neutrons emitted during a random fission

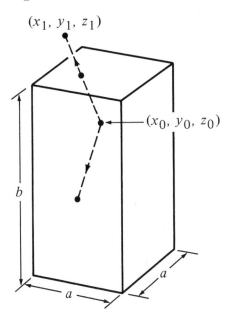

(x_1, y_1, z_1)

(x_0, y_0, z_0)

b

a

a

then we must choose random values for the coordinates x_0, y_0, and z_0, subject to the conditions

$$-\frac{a}{2} < x_0 < +\frac{a}{2}$$

$$-\frac{a}{2} < y_0 < +\frac{a}{2}$$

$$-\frac{b}{2} < z_0 < +\frac{b}{2}.$$

The only fission fragments that we are concerned with are the neutrons, since the heavy nuclear fragments play no part in the chain reaction mechanism. The two neutrons emitted during the fission process may travel in any directions. (We shall ignore the fact that the number of emitted neutrons is not always two; and we shall also ignore possible correlations between the two neutron directions.) A direction in three dimensions can be specified by two angles: θ, the polar angle, and ϕ, the azimuth (see Figure 4.13). If the emitted neutrons have an "isotropic" distribution, i.e., all directions are equally likely, then the probability of a neutron emitted from the point (x_0, y_0, z_0) hitting any area on a surrounding unit sphere depends only on the size of the area. This implies that the azimuth ϕ is uniformly distributed between 0 and 2π, and that $\cos\theta$ is uniformly distributed between -1 and $+1$. Note that it is $\cos\theta$, and not θ, which is uniformly distributed, due to the fact that equal intervals in $\cos\theta$ contain the same area on a sphere of unit radius.

Whether an emitted neutron hits another nucleus before leaving the block depends only on the distance along its line of flight to the boundary of the block. We shall assume (unrealistically) that a neutron emitted during fission can hit another nucleus after it travels any distance between 0 and 1 centimeters, with equal probability. For example, if a neutron travels along a direction such that

Figure 4.13
A neutron emitted at the point (x_0, y_0, z_0) travels along a direction specified by the angles θ and ϕ.

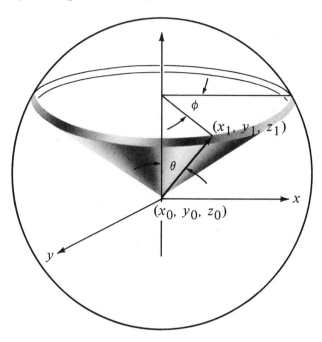

it would leave the block after traveling only 0.3 cm, then there is a 30% chance of it hitting a nucleus in the block. Our procedure, therefore, is to choose a random number between 0 and 1 for d, the distance traveled by each neutron. Since the neutron starts at the point (x_0, y_0, z_0) and travels along a direction (θ, ϕ), we can find the coordinates of the point (x_1, y_1, z_1) where the neutron would hit another nucleus using the geometrical relations:

$$x_1 = x_0 + d \sin \theta \cos \phi$$
$$y_1 = y_0 + d \sin \theta \sin \phi \qquad (4.8)$$
$$z_1 = z_0 + d \cos \theta.$$

Whether the neutron actually hits a nucleus at the point (x_1, y_1, z_1) and causes it to fission depends on whether the point lies within the block. If we count the total number of times (N_{in}) that the neutron endpoints (x_1, y_1, z_1) lie within the block for all N random fissions, we can compute the survival fraction f, using $f = N_{in}/N$.

Program to Calculate the Survival Fraction f

The program shown in flow diagram form in Figure 4.14, and listed on pages 205-207, calculates the survival fraction f for a block of uranium of specified size and shape. The program first reads from a data card numerical values for the parameters M, S, and N, where

    M = mass of the uranium block
    S = a/b (the ratio of the lengths of two of the sides of the block)
    N = number of random fissions to use in calculating f.

The program then finds the dimensions of the block (a and b) from the values of M and S. (The reason for reading values for M and S instead of a and b directly is to make it easier to study the dependence of the survival fraction f on the mass and shape of the block separately.) We assume that the units are such that the density of the block is equal to one, so that the mass and volume of the block are numerically equal. We therefore have

$$M = V = a^2b = \frac{a^3}{S} = b^3S^2, \tag{4.9}$$

where $S = a/b$. Using equation 4.9, we can find a and b from M and S:

$$a = (MS)^{1/3} \qquad b = \left(\frac{M}{S^2}\right)^{1/3}.$$

The program proceeds to generate N random fissions, as explained in the previous section. To generate each random fission, we need nine random numbers $r_1, r_2, \ldots, r_9$, which lie between 0 and 1. (The random number generator was discussed in section 4.1.) The nine quantities needed for each random fission are obtained from the random numbers $r_1, r_2, \ldots, r_9$ according to the following equations:

$$\left.\begin{array}{l} x_0 = a(r_1 - \frac{1}{2}) \\ y_0 = a(r_2 - \frac{1}{2}) \\ z_0 = b(r_3 - \frac{1}{2}) \end{array}\right\} \quad \text{coordinates of the nucleus undergoing fission}$$

$$\left.\begin{array}{l} \phi = 2\pi r_4 \\ \cos\theta = 2(r_5 - \frac{1}{2}) \end{array}\right\} \quad \text{two angles for one emitted neutron}$$

$$\left.\begin{array}{l} \phi' = 2\pi r_6 \\ \cos\theta' = 2(r_7 - \frac{1}{2}) \end{array}\right\} \quad \text{two angles for the other emitted neutron}$$

$$\left.\begin{array}{l} d = r_8 \\ d' = r_9 \end{array}\right\} \quad \text{distances traveled by each neutron}$$

These formulas give the proper range for each of the nine parameters. The program computes the neutron endpoint using equation 4.8 and determines whether the point (x_1, y_1, z_1) lies within the block, in which case it adds 1 to the value of N_{in}, the number of induced fissions. After generating N random fissions, it computes the survival fraction $f = N_{in}/N$ and prints the result.

Comments on Data Cards and Sample Results

The output shown in Figure 4.15 was obtained using three data cards containing the following values:

	M	S	N
1st card	1.0	1.0	100.0
2nd card	1.0	1.0	100.0
3rd card	1.0	2.0	100.0

Different values for f when identical values for the parameters M, S, and N are used is indicative of the statistical fluctuations present in all Monte Carlo calculations (see output for first two cards). The accuracy of the calculation can be increased by using a larger value for N.

The use of the Monte Carlo method in this problem is actually quite similar to the first application (the evaluation of integrals to find the mass and center of

Figure 4.14   Flow diagram to calculate the survival fraction f

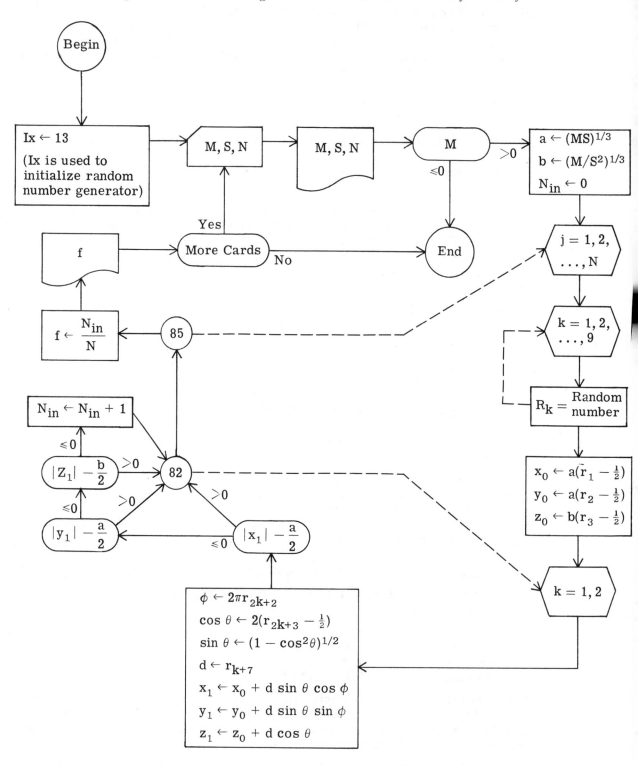

mass coordinates of an object.) In finding the survival fraction f, we are actually integrating a function F of nine variables:

$$F = F(x_0, y_0, z_0, \phi, \theta, d, \phi', \theta', d'),$$

which represents the number of fissions induced by two emitted neutrons for particular values of the variables $x_0, y_0, \ldots, d'$. The value of F is zero, one, or two, depending on these variables. In order to obtain the survival fraction f, we, must integrate the function F over all nine variables. The advantages of the Monte Carlo technique over conventional integration techniques are quite apparent in a case such as this. In order to evaluate a nine-dimensional integral using a finite sum, each of the nine variables must be allowed to take on some number of values. If only two values are used for each variable, it becomes necessary to evaluate the function at $2^9 = 512$ points.

Figure 4.15

```
M=    1.0000  S=    1.0000  N=  100.0000  F=    .3100
M=    1.0000  S=    1.0000  N=  100.0000  F=    .2800
M=    1.0000  S=    2.0000  N=  100.0000  F=    .2100
M=     .0000  S=     .0000  N=     .0000
```

Problems 12-14 at the end of this chapter pertain to the simulation of a chain reaction.

```
C           CALCULATION OF THE CRITICAL MASS FOR A FISSION CHAIN REACTION
C
C---------------------------------------------------------------------------
C
C           THIS PROGRAM COMPUTES THE SURVIVAL FRACTION F, FOR A
C           RECTANGULAR SLAB OF FISSIONABLE MATERIAL.  F IS DEFINED AS THE
C           NUMBER OF INDUCED FISSIONS RESULTING FROM ONE SPONTANEOUS FISSION.
C           FOR VALUES OF F GREATER THAN 1.0, A CHAIN REACTION OCCURS.  THE
C           SLAB HAS THE 'CRITICAL MASS' WHEN F = 1.
C           WE ASSUME THAT TWO NEUTRONS ARE EMITTED DURING EACH FISSION.
C
      REAL M,NIN,N
      DIMENSION R(9)
C
C           IX IS SET TO AN ODD INTEGER TO INITIALIZE THE RANDOM NUMBER
C           GENERATOR.
C
C
      PI=3.14159265
      IX=13
C           READ A DATA CARD CONTAINING M, S, AND N.
C           M IS THE MASS OF THE RECTANGULAR SLAB.
C           S IS THE SHAPE OF THE SLAB (THE RATIO OF THE LENGTH  TO THICKNESS)
C           N IS THE NUMBER OF RANDOMLY GENERATED NEUTRONS
C
   10 READ(2,1000)M,S,N
      WRITE(3,1001)M,S,N
C
C           IF THE DATA CARD READ IS BLANK, M IS ZERO AND WE QUIT.
C
      IF(M)90,90,20
C
```

```
C                  FIND THE DIMENSIONS OF THE SLAB FROM M AND S.
C
   20  A=(M*S)**1.0/3.0
       B=(M/S**2)**1.0/3.0
C
C                  SET NIN, THE NUMBER OF CASES IN WHICH AN EMITTED NEUTRON REACTS
C                  WITH A NUCLEUS INSIDE THE BOUNDARIES OF THE SLAB, EQUAL TO ZERO
C                  INITIALLY.
C
       NIN=0.0
       NUM=N
C
C                  THIS LOOP GENERATES N RANDOM FISSIONS, WHICH GIVE RISE TO
C                  2N RANDOM NEUTRONS.
C
       DO 85 J=1,NUM
C
C                  FOR EACH RANDOM FISSION WE NEED  9 RANDOM NUMBERS.
C
       DO 50 K=1,9
C
C                  THE INSTRUCTIONS INCLUDED BELOW BETWEEN THE LINES GENERATE
C                  A RANDOM NUMBER, RAN.
C
C-----------------------------------------------------------------------
       IX=IABS(899*IX)
       IF(IX-32767)5002,5002,5001
 5001  IX=IX-(32767+1)*(IX/(32767+1))
 5002  RAN=IX
       RAN=RAN/32767.
C-----------------------------------------------------------------------
   50  R(K)=RAN
C
C                  THE X, Y, AND Z COORDINATES OF THE NUCLEUS EMITTING A NEUTRON
C                  ARE RANDOM NUMBERS INSIDE THE BOUNDARIES OF THE BLOCK.
C                  X AND Y RANGE FROM -A/2 TO +A/2, AND Z RANGES FROM -B/2 TO +B/2.
C
       X0=A*(R(1)-0.5)
       Y0=A*(R(2)-0.5)
       Z0=B*(R(3)-0.5)
C
C                  WE NEED TWO RANDOM NEUTRONS FOR EACH FISSION.
C
       DO 82 K=1,2
C
C
C                  TWO ANGLES ARE NEEDED TO DEFINE A DIRECTION IN SPACE
C
       PHI=2.0*PI*R(2*K+2)
       COSTH=2.0*(R(2*K+3)-0.5)
       SINTH=SQRT(1.0-COSTH**2)
C
C                  THE DISTANCE TRAVELED BY A NEUTRON IS ASSUMED TO BE A RANDOM
C                  NUMBER FROM 0 TO 1.
C
       D=R(K+7)
C
C                  FIND THE COORDINATES OF THE INTERACTION POINT FOR THE NEUTRON.
C
       X1=X0+D*SINTH*COS(PHI)
       Y1=Y0+D*SINTH*SIN(PHI)
```

```
         Z1=Z0+D*COSTH
C
C              IF NEUTRON INTERACTION POINT IS INSIDE THE BLOCK, ADD 1 TO NIN.
C
         IF(ABS(X1)-A/2.)60,60,82
   60    IF(ABS(Y1)-A/2.)70,70,82
   70    IF(ABS(Z1)-B/2.)80,80,82
   80    NIN=NIN+1.0
C
C              DO NEXT NEUTRON FOR THIS FISSION.
C
   82    CONTINUE
C
C              DO NEXT RANDOM FISSION.
C
   85    CONTINUE
C
C              THE SURVIVAL FRACTION IS DEFINED AS NIN/N.
C
         F=NIN/N
         WRITE(3,1002)F
C
C              GO READ NEXT DATA CARD.
C
         GO TO 10
 1000    FORMAT(3F10.5)
 1001    FORMAT(4H  M=,F10.4,4H  S=,F10.4,4H  N=,F10.4)
 1002    FORMAT(1H+, 42X,4H  F=,F10.4)
   90    CONTINUE
         CALL EXIT
         END
```

4.4   The Approach to Equilibrium

The second law of thermodynamics states that the entropy of an isolated system must either stay constant or increase. In other words, when things are left alone, they either "coast" or "run down," i.e., they tend to approach a steady equilibrium state. This principle, which conforms to our everyday observations of complex systems, appears to be contradicted by the behavior of simple systems which contain a small number of particles. To illustrate the difficulty, let us consider a box partitioned in the middle with a number of gas molecules all initially to the left of the partition. If a small hole is made in the partition, some molecules enter the right half of the box. If the number of molecules is very large, as is ordinarily the case with a gas in a macroscopic box, then within the limits of accuracy of our instruments, we always find that the system tends to approach equilibrium, which in this case means equal pressures, or equal numbers of molecules in each half of the box. (See Figure 4.16a.) The reverse process (starting with equal numbers of molecules in each half and later finding all in one half) is never observed. Thus, we can infer with (near) certainty the time sequence of the pictures in Figure 4.16a from the pictures themselves. However, if the box contained a small number of molecules we could no longer determine the time sequence from the pictures themselves. If the left half of the box initially contains six gas molecules, then due to their random motion we would find that at some future time after the hole is made, the system has reached the equilibrium state of three molecules in each half of the box. However, it is not at all unlikely that at some still later time, due to the random molecular motion, the initial state of all six molecules in the left half of the box will recur making it impossible to infer the time sequence of the pictures in Figure 4.16b from the pictures themselves. In fact, using statistics we could determine how long, on the average, we would have to wait for such a recurrence of the initial state. If a large number of molecules are in the box, such a recurrence of the initial state is still expected, but only after a fantastically long time. Over any observable time (say one human lifetime), large departures from the equilibrium state, once it is reached, occur with vanishingly small probability. Thus, the thermodynamic principle that the entropy of an isolated system does not decrease, or that the system can approach equilibrium but not depart from it, must be understood on a statistical basis— this is what (almost) always happens to systems of very many particles over an observable period of time. Over a very long period of time, however, an isolated system can depart from its equilibrium state by an arbitrarily large amount, the magnitude of the departures from equilibrium being more readily observable the longer the time period, and the fewer the number of particles in the system.

Figure 4.16
The tendency toward equilibrium for molecules in a partitioned box

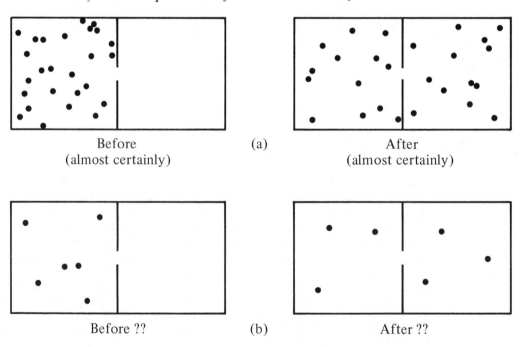

| Before | (a) | After |
| (almost certainly) | | (almost certainly) |

| Before ?? | (b) | After ?? |

A Simulation of the Approach to Equilibrium Using N Coins

To study the time evolution of the N particles in a box we use a model involving N coins. A coin which is face up (heads) represents a particle in the left half of the box, and one which is face down (tails) a particle in the right half. To represent a particle going through the hole in the partition we turn over a coin selected at random. This assumes that it is purely a matter of chance which of the particles numbered, 1, 2, ..., N, goes through the hole. Starting with all coins heads (all particles in the left half of the box), we can follow the evolution of the N-particle gas by repeatedly choosing a coin at random and turning it over. Since all the coins are initially heads, if N is large, it is likely that the first few turns will result in an increase in the number of tails. In general, on any turn of a randomly chosen coin, the chances of an increase in the number of tails depends only on the fraction of all coins that are heads. Thus, the tendency for the system to reach the equilibrium state of equal numbers of heads and tails (equal numbers of particles in the two halves of the box), is simply a consequence of the laws of statistics.

Let us use statistical considerations to determine the manner in which this system approaches equilibrium. We let n designate the number of tails after x randomly chosen coins have been turned over. On the next turn, the probability p_h of a coin being turned from tails to heads is just the fraction of all coins that are tails: $p_h = f = n/N$. The probability of a coin being turned from heads to tails is therefore $p_t = 1 - p_h$. Hence, on the average, the change in the number of tails on the next turn is given by

$$\Delta n = p_t - p_h = 1 - \frac{2n}{N}.$$

If Δx randomly chosen coins are turned over, instead of just one, we have

$$\Delta n = \left(1 - \frac{2n}{N}\right)\Delta x. \tag{4.10}$$

Treating n and x as continuous variables, and using differentials instead of finite differences, we divide both sides of equation 4.10 by the factor $(1 - 2n/N)$ to obtain

$$\frac{dn}{\left(1 - \dfrac{2n}{N}\right)} = dx$$

which can be integrated to give

$$-\frac{N}{2} \ln (N - 2n) = x + C.$$

We choose the constant of integration C equal to $-\frac{1}{2}N \ln N$, in order to conform to the assumed initial condition n = 0 for x = 0 (all coins initially heads). Solving for the fraction $f = n/N$, we obtain

$$f = n/N = \tfrac{1}{2}(1 - e^{-2x/N}). \tag{4.11}$$

According to equation 4.11, the fraction of coins which are tails, *on the average*, after x coins have been turned over is an exponential function of x which approaches the equilibrium value of $\frac{1}{2}$ for $x \to \infty$. If we were to plot the fraction of coins that are tails, $f = n/N$ versus x, based on the results of an actual coin turning experiment, we would find an irregular curve resembling, on the average, an exponential function with random statistical fluctuations superimposed. The magnitude of the fluctuations decrease as the number of coins increases. For a small number of coins (or molecules) the relatively large fluctuations may completely obscure any exponential approach to equilibrium, and in fact make the concept somewhat meaningless.

Figure 4.17   *Flow diagram for the approach to equilibrium*

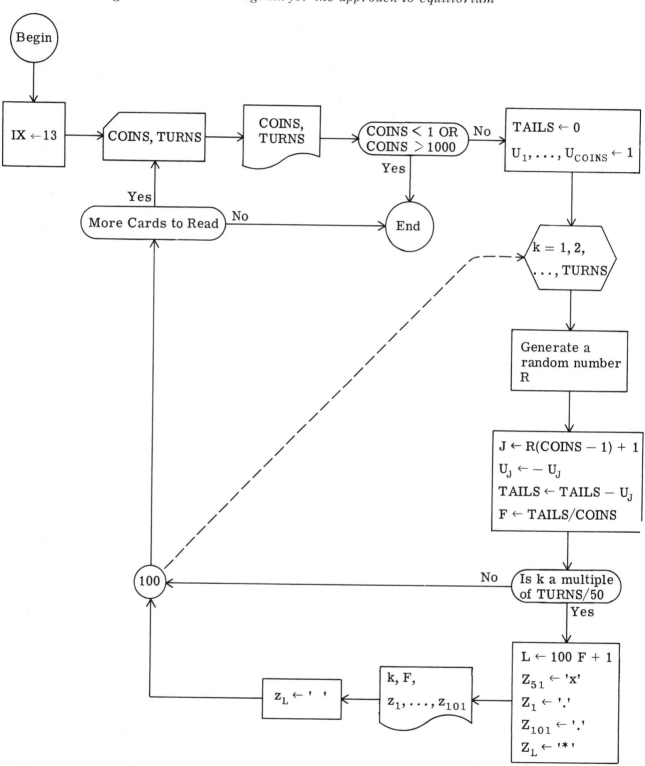

Program to Simulate the Approach to Equilibrium

The program shown in flow diagram form in Figure 4.17 and listed on pages 213-214 simulates an approach to equilibrium using the coin model discussed above. The program begins by reading from a data card numerical values for COINS and TURNS, where

COINS = number of coins
TURNS = number of times to turn over coins selected at random.

The array $U_1, \ldots, U_{1000}$, is used to record the state of each of up to 1000 coins according to the convention:

$U_j = +1$ means the jth coin is heads
$U_j = -1$ means the jth coin is tails

Figure 4.18

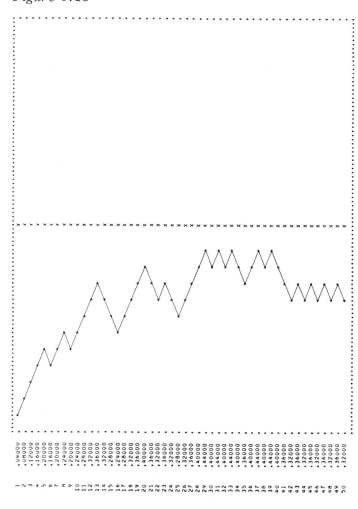

Figure 4.19

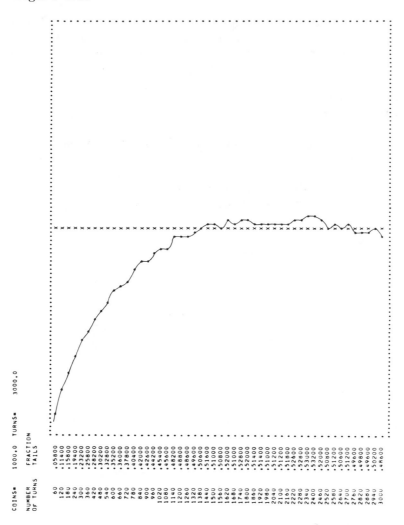

The program initially sets all the elements of the array U equal to +1 (all coins initially heads). Then in a loop over $k = 1, 2, \ldots,$ TURNS, an integer j is chosen at random between 1 and COINS, and the jth coin is turned over by simply reversing the sign of U_j. The number of tails is stored in the variable TAILS, so that by subtracting U_j from TAILS each time a coin is turned the program correctly accounts for changes in the number of tails. (It increases the number of tails by one if the coin is turned from heads to tails and decreases it by one in the opposite case.) The program computes the fraction of coins that are tails, prints and plots the results provided that this turn is one of 50 selected turns evenly interspersed between 1 and TURNS. (The reason for only printing the results at 50 selected turns is to get a single-page plot. Because of this, you should choose values for TURNS that are integral multiples of 50.) Once the loop over the specified number of turns is complete, the program goes on to read another data card if any remain to be read.

Comments on Data Cards and Sample Results

The output shown in Figures 4.18-4.19 was produced using two data cards containing the following numerical values for the parameters:

	COINS	TURNS
1st card	25.0	50.0
2nd card	1000.0	3000.0

The random statistical fluctuations are much more pronounced when the number of coins is small, as in the first case. Note that when a large value for COINS is used, it is also necessary to use a large value for TURNS if we wish to reach the equilibrium state. The value used for COINS on the second data card (1000) is the largest allowed value due to the size of the array U. While there is no such upper limit on the value of TURNS, the amount of computation time will depend on this value. Recall that the value chosen for TURNS should be an integral multiple of 50.

Problems 15-19 at the end of this chapter pertain to the approach to equilibrium.

```
C             APPROACH TO EQUILIBRIUM
C
C---------------------------------------------------------------------------
C
C             THIS PROGRAM CALCULATES AND PLOTS THE FRACTION OF COINS IN A SET
C             OF COINS THAT ARE TAILS AFTER A NUMBER OF RANDOM TURNS ARE
C             MADE.  IT IS ASSUMED THAT ALL COINS ARE INITIALLY HEADS.
C
      DIMENSION U(1000),Z(101)
      DATA EX, ASTER, BLANK, DOT/'X','*',' ','.'/
C
C             IX IS NEEDED TO INITIALIZE THE RANDOM NUMBER GENERATOR
C
      IX=13
C
C             READ A DATA CARD CONTAINING COINS AND TURNS.
C             COINS = NUMBER OF COINS
C             TURNS = NUMBER OF RANDOM TURNS
C
   10 READ(2,1000)COINS,TURNS
      WRITE(3,1001)COINS,TURNS
      NC=COINS
      NT=TURNS
      NS=NT/50
C
C             IF THE NUMBER OF COINS IS EITHER ZERO OR GREATER THAN 1000, QUIT.
C
      IF(ABS(COINS-500.5)-500.0)20,20,110
C
C             THE NUMBER OF TAILS IS INITIALLY ZERO.
C
   20 TAILS=0.
C
C             INITIALLY, WE ASSUME ALL THE COINS ARE HEADS.  U(1),....,U(M)=+1.
C
      DO 30 K=1,NC
   30 U(K)=+1.0
C
C             SET Z(1), Z(2), Z(3),...., Z(101) ALL EQUAL TO THE CHARACTER BLANK.
C
      DO 40 K=1,101
   40 Z(K)=BLANK
```

```
      WRITE(3,1003)
      DO 100 K=1,NT
C
C              GENERATE A RANDOM NUMBER R.
C
C----------------------------------------------------------------
C
      IX=IABS(899*IX)
      IF(IX-32767)50,60,60
  50  IX=IX-(32767+1)*(IX/(32767+1))
  60  CONTINUE
      RAN=IX
      RAN=RAN/32767.
C
C----------------------------------------------------------------
C
C              J IS A RANDOM INTEGER FROM 1 TO NC.
C
      J = R*(COINS-1.0)+1.0
C
C              TURN JTH COIN OVER.
C
      U(J)=-U(J)
C
C              REVISE THE NUMBER OF TAILS.
C
      TAILS=TAILS-U(J)
      F=TAILS/COINS
C
C              ONLY DISPLAY RESULTS IF K IS A MULTIPLE OF NS = TURNS/50.
C
      IF(K-NS*(K/NS))95,95,100
  95  L=100.*F+1.0
      Z(51)=EX
      Z(1)=DOT
      Z(101)=DOT
      Z(L)=ASTER
      WRITE(3,1002)K,F,(Z(J),J=1,101)
      Z(L)=BLANK
 100  CONTINUE
      WRITE(3,1003)
      GO TO 10
1000  FORMAT(2F10.5)
1001  FORMAT(7H1COINS=,F10.1,8H  TURNS=,F10.1,          //20H NUMB
     1ER    FRACTION/20H OF TURNS    TAILS    /)
1002  FORMAT(1X,I6,F10.5,2X,101A1)
1003  FORMAT(19X,101H.................................................
     1............................................)
 110  CONTINUE
      CALL EXIT
      END
```

Problems for Chapter 4

Program to Generate Pseudorandom Numbers

1. Length of period.   Modify the program so that it counts how many numbers are generated before the first number in the sequence, x_0, repeats. Run the modified version using a number of values for the parameters N, c, and x_0. Compare your findings with the assertions on page 179: (a) For given values of c and N, odd values of x_0 give a period which is twice that of even values; (b) For given values of x_0 and N, the longest period is obtained using values of c which satisfy $c = 8n \pm 3$, where n is any positive integer; (c) When x_0 and c satisfy the conditions specified in (a) and (b), the length of the period is N/4. You should only use values for N given by $N = 2^b$ (for a binary computer), or $N = 10^d$ (for a decimal computer). When you modify the program to find the period, you should also suppress the printout of individual random numbers

2. Frequency distribution test.   Modify the program so that it generates a sequence of pseudorandom numbers and keeps a count of how many numbers lie in each of ten equal subintervals from zero to one. Use $N = 32768.$, $X = 13.$, $C = 899.$ for the parameters. Make a number of runs of 300 numbers each, and compare the rms deviations from a uniform distribution with the expected deviations predicted according to the laws of statistics. Repeat the procedure using another value of C, perhaps $C = 5$. Does the frequency distribution test give satisfactory results for both values of C?

3. Frequency of runs test.   Modify the program so that it generates a sequence of pseudorandom numbers and keeps a count of the number of runs of different lengths. Compare the results you obtain with the expected results for a truly random sequence of the same length. Do this for both type of runs discussed on page 183, i.e., the string of monotonically increasing or decreasing numbers, and the string of numbers all above, or all below, the value $\frac{1}{2}$. Use the following values of the parameters: $N = 32768.$, $C = 899.$, $X = 13$. Repeat the whole procedure using another value of C, perhaps $C = 5$. Does the frequency of runs tests give satisfactory results for both values of C?

4. Nonuniformly distributed pseudorandom numbers.   Modify the program so that it generates a sequence of pseudorandom numbers which are distributed according to some nonuniform distribution f(x), using the procedure outlined on page 183. Run the program using $f(x) = x$ (for $0 < x < 1$). Have the program keep a count of the number of numbers lying in each of ten subintervals from zero to one, in order to verify that the actual distribution of numbers conforms to $f(x) = x$. Try using other functions for f(x).

Monte Carlo Evaluation of Definite Integrals

5. Variation of the calculated mass and center of mass coordinates for a fixed N. Run the program using a data card with $N = 100$ and NRUNS = 50. Make a histogram showing the number of runs that the calculated mass lies in each of a number of equal-size intervals. Do the same for the calculated x and y center of mass coordinates, combining the values for both coordinates in one plot, instead of making two separate plots. Both histograms should have a shape which resembles a Gaussian distribution, and the latter histogram should be centered near zero, if the random number generator is working properly.

6. Variation of the results with N.   Repeat the procedure suggested in the preceding problem using other values of N, say $N = 25$ and $N = 400$. Compare the distributions you obtain in these two cases with the original ones. In particular, see if the predicted $N^{-1/2}$ dependence of the error in each quantity (determined from the width of the distribution) appears to be valid.

7. *Objects of other shapes*.   Modify the program so that it computes the mass
and center of mass coordinates for objects of a variety of shapes. For example,
you can try the shapes shown in Figure 4.20. Be sure to select random numbers
for each coordinate between limits which define a rectangular block that just
encloses the object. For each shape, use a large enough N to get a reasonably
accurate result, and get an estimate of the error by using NRUNS > 1. Whenever possible compare the results with theoretical predictions.

Figure 4.20

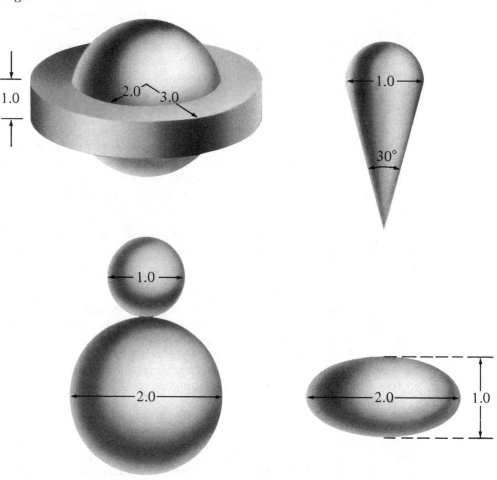

8. *Nonuniform density*.   Modify the program so that it computes the mass and
center of mass coordinates of a spherical planet of radius $R_0 = 10^6$ meters,
whose density is given by the formula

$$\rho(x, y, z) = 1 + e^{-R/R_0},$$

where $R = (x^2 + y^2 + z^2)^{1/2}$.

9. Moment of inertia.   The moment of inertia of a continuous body about the j-axis is defined as

$$I_j = \int r_j{}^2 dm.$$

If we replace the continuous body by a large number of randomly located points within the body, this becomes

$$I_j = \sum_{j=1}^{N} r_j{}^2 \Delta m_j.$$

Thus, the moments of inertia of the body about each of the three coordinate axes are given by

$$I_x = \sum x_j{}^2 \Delta m_j$$

$$I_y = \sum y_j{}^2 \Delta m_j$$

$$I_z = \sum z_j{}^2 \Delta m_j.$$

Modify the program so that it computes the moments of inertia for objects of varied sizes, densities, and shapes. In particular, try a sphere, a cube, and a disk, and compare your results with theoretical predictions.

10. Integrals of more than three dimensions.   Integrals of more than three dimensions can be calculated using a Monte Carlo technique in which the randomly located points have more than three coordinates. For example, we can find the mass of a *hypersphere,* the four-dimensional analog of a sphere, defined by the equation

$$w^2 + x^2 + y^2 + z^2 = R_0^2,$$

by generating a large number of points (w, x, y, z) having random locations within a hypercube one unit on a side, and counting how many points lie inside the hypersphere (assumed to have a radius $R_0 = 0.5$). A point lies inside the hypersphere if

$$w^2 + x^2 + y^2 + z^2 < (.5)^2.$$

Modify the program to calculate the mass and center of mass coordinates of a hypersphere.

11. Change of variables.   There are many problems in which the previously discussed Monte Carlo technique for finding the mass and center of mass coordinates of an object can be improved using a change of variables. For example, suppose we wish to find the mass and center of mass coordinates of a *thin* spherical shell of radius R. The shell has a *surface* mass density σ, which may be a function of position on the shell. If we were to simply enclose the shell inside a cube and then pick points at random inside the cube, then very few of these points would lie inside the spherical shell. Thus, the statistical fluctuations would be very large, unless an enormous number of points were used. A much better approach is to change variables so as to pick *only* random points which lie in the thin spherical shell. We can then compute the mass of the sphere using

$$M = \sum_{j=1}^{N} \Delta m_j, \qquad (4.12)$$

where Δm_j, the mass associated with each random point, is given by

$$\Delta m_j = \sigma_j \Delta a, \qquad (4.13)$$

where σ_j is the surface mass density of the shell at the location of the jth point, and Δa is the average surface area per mass point on the shell, given by

$$\Delta a = \frac{4\pi R^2}{N}. \qquad (4.14)$$

In order to generate points which lie at random locations on the surface of a sphere of radius R, we need to choose values for the two angles θ and ϕ (see Figure 4.13 on page 202. In order that equal areas of the sphere contain equal numbers of random points, we need to obtain the angles θ and ϕ using

$$\phi = 2\pi r_1$$
$$\cos \theta = 2(r_2 - \tfrac{1}{2}),$$

where r_1 and r_2 are random numbers having a uniform distribution between zero and one.

If the surface mass density σ is a known function of the variables $\cos \theta$ and ϕ, the mass of each random point can be computed using equations 4.13 and 4.14, and the total mass of the shell can then be found using equation 4.12. Write a program to carry out the procedure discussed above. Run the program to find the mass of a thin spherical shell of unit radius, whose surface mass density is given by $\sigma = 1 + \tfrac{1}{2} \cos^2 \theta$. (As a check on the program, you might first want to use $\sigma = 1$.)

Program to Simulate a Chain Reaction

12. Calculation of the critical mass.   Determine the critical mass for a cube $(a/b = 1.0)$ by running the program using $S = 1.0$, and a range of values for the mass of the block M and the number of random fissions N. The critical mass is that mass for which the computed survival fraction f is equal to 1.0. A good strategy might be to use a relatively small value for N, say $N = 100$, until the computed values of f are in the vicinity of 1.0, and then use larger values for N to attain greater accuracy.

By repeating the above procedure, you can determine the critical mass for non-cubic rectangular blocks $(S \neq 1.0)$. Find the critical mass using a number of values of S both less than one and greater than one. Explain the dependence you observe of the critical mass on the parameter $S = a/b$.

13. Calculations of the critical mass for a spherical piece of uranium.   Modify the program so that the random fissions occur inside a spherical piece of uranium. Use the same technique as in the present version of the program to calculate the survival fraction f. Run the modified version of the program to calculate f for a range of masses, and determine the critical mass for a sphere. Compare the result with what you found for the rectangular block.

14. Minimum separation distance between two subcritical blocks.   Modify the program so that it generates random fissions within two nearby blocks of uranium separated by a distance s. In counting the number (N_{in}) of induced fissions we need to determine if the neutron endpoints from random fissions lie within *either* block. Run the modified version of the program using two cubic blocks of uranium, each of which is 90% the critical mass. (You should have found the critical mass for a cube (S = 1.0) in problem 12.) Use a range of values for s, the distance between blocks, which can be read from a series of data cards. Find the separation between blocks below which a chain reaction would occur (f = 1.0).

Program to Simulate the Approach to Equilibrium

15. The approach to equilibrium for a large number of coins.   Run the program using values for COINS in the range 25 to 1000, with appropriate values for TURNS. Since any single approach to equilibrium is subject to random statistical fluctuations, you should try several runs for each value of COINS and TURNS. Using the results, you can check the predicted exponential approach to equilibrium (equation 4.11), which should hold *on the average* (i.e., apart from statistical fluctuations). One simple check to make is that according to equation 4.11, for $x = \frac{1}{2}N$, we should have $f = \frac{1}{2}(1 - e^{-1}) \approx 0.316$.

16. Recurrence of the initial state for a small number of coins.   Run the program using values for COINS of 3, 4, 5 and 6, and TURNS = 50. For numbers of coins this small, the large random statistical fluctuations make it meaningless to speak of an approach to equilibrium. According to the laws of statistics, the initial state (all coins heads) should recur, on the average, at intervals of 2^N turns, where N = COINS. Check this prediction for the cases of 3, 4, 5, and 6 coins. (Note that when the number of coins is reasonably large, say N = 100, a recurrence of the initial state is still predicted, but only on the average of once every 2^{100} turns. The size of this number is so great that if we could turn random coins over a billion times a second, we would have to wait, on the average, ten thousand times the present age of the universe (about ten billion years) before seeing a recurrence of the initial state, after equilibrium has been reached.)

17. Deviations from equilibrium.   Modify the program so that it computes the number of times the value of f = n/N (the fraction of coins that are tails) falls in each of a series of intervals centered about the equilibrium value f = 0.5. A reasonable choice would be 50 equal-size intervals from 0.4 to 0.6. In order that the program only deal with departures from the equilibrium state after equilibrium has been reached, it should not begin counting the number of cases in each interval until after a number of random flips equal to twice the number of coins. Run the modified version of the program using values of the number of coins from 25 to 1000 and make histograms from the results showing the relative frequency of deviations from equilibrium for each case. Discuss the shape of each histogram, and the dependence of the width of each distribution on the values chosen for COINS and TURNS.

18. Actual experiment using coins.   Using a set of from 25 to 50 coins, perform an actual experiment. Starting with all heads, turn over randomly selected coins, each turn recording the fraction of coins that are tails. Plot your results and compare them with the computer simulation for the same number of coins. Be sure that you take adequate precautions to select coins in a completely random manner—for instance, keep your eyes closed and withdraw your hand each time you have turned over a coin, so that there is a chance of the next coin being the same as the previous one.

19. Radioactive disintegration.   The same model using N coins can be used to simulate the disintegration of a large number of nuclei. Let a coin which is heads represent a nucleus that has not yet disintegrated. The probability that any one nucleus disintegrates in any short time interval Δt is a constant p_0, independent of how long the nucleus has been in existence.

To simulate the disintegration process, we start with all coins heads (no nucleus yet disintegrated). We set up a loop to step the time: $t = 0, \Delta t, 2\Delta t, \ldots,$ $n\Delta t$. For each value of the time, we set up a loop over the N nuclei, $j = 1, 2, \ldots,$ N. For each nucleus, we test whether it has already disintegrated (jth coin tails), and if so, we skip this one and go on to the next. If the jth nucleus has not yet disintegrated, we "give it a chance" to disintegrate by generating a random number r, uniformly distributed from zero to one, and having the nucleus disintegrate if r is less than p_0. When the nucleus disintegrates, we turn the jth coin over. After giving all N nuclei a chance to disintegrate for a particular time interval, we compute the fraction of nuclei that have not yet disintegrated f, and print out both its numerical value, along with a line of characters for a plot of f versus time. We repeat the process for each time interval, until all n time intervals have been treated (or until all N nuclei have disintegrated).

Modify the original (approach to equilibrium) program, so that it uses the procedure outlined above to simulate the process of radioactive disintegration. Run the program using a number of values for the parameters:

> N = Number of nuclei present at time $t = 0$
> Δt = time step
> n = number of time steps
> p_0 = probability that a nucleus disintegrates in a time interval Δt.

Note that the probability p_0 is related to the nuclear *lifetime*, T, which appears in the formula for the predicted exponential time dependence of f versus time:

> $f(t) = e^{-t/T}$,

which should hold for large values of N. According to this formula after a time t equal to the lifetime T, the fraction of nuclei which have not yet disintegrated is 1/e. The relation between the probability p_0 and the lifetime T is

> $$p_0 = \frac{\Delta t}{T}.$$

5

Analysis of Experimental Data

Computers can be used in a variety of ways related to the analysis of experimental data. In this chapter we discuss several problems which illustrate two particular computer uses—least squares fitting of experimental data and the calculation of physical quantities from the data.

5.1 Least Squares Fitting of Experimental Data

In analyzing data from an experiment, we generally attempt to determine if the measured variables satisfy some particular mathematical relationship. Suppose the data consists of a set of n values of some measured quantity y: $y_1, y_2, \ldots, y_n$, corresponding to n associated values of some variable x: $x_1, x_2, \ldots, x_n$. In most experiments it is also possible to assign to each data point (x_j, y_j), a random error of measurement, Δy_j, which depends on the precision of the measuring apparatus. There may also be an uncertainty in the x variable, but we shall simplify the analysis by ignoring this possibility.

Suppose we wish to test whether x and y satisfy some particular functional relationship $y = f(x)$. The simplest test of whether the data points are consistent with the hypothesized relation is to plot the function $y = f(x)$ on the same graph as the data points $(x_1, y_1), \ldots, (x_n, y_n)$. The data points are considered consistent with the function $y = f(x)$ if they lie "near" the curve and are scattered on either side of it in a random manner. It is highly unlikely that all the points lie exactly on the curve, due to measurement error. For example, the data points in Figure 5.1 would be considered quite consistent with the function represented by the curve. The vertical *error bars* on each point extend above and below the point a distance equal to the measurement error Δy_j. A data point could be considered "near" to the curve if its distance to the curve in the y direction is no greater than the assigned measurement error Δy_j, making the error bar intersect the curve. This vertical distance from the data point to the curve is the *residual* for the point (x_j, y_j), and is given by

$$r_j = y_j - f(x_j). \tag{5.1}$$

Note that the residual r_j is positive or negative, depending on whether the data point is above or below the curve.

We might consider a set of data consistent with a curve even if the magnitudes of some of the residuals r_j are greater than the corresponding measurement errors Δy_j, which is the case for two of the data points in Figure 5.1. This is because the measurement error Δy_j is usually defined such that there is some specific probability (less than 100%) of any particular measurement falling within a range of $\pm \Delta y_j$ of the true value. In fact, one way to experimentally determine the measurement error Δy_j is to make many repeated measurements of y_j for the *same* value of x_j, and then find a value for Δy_j such that a specific

Figure 5.1
A set of data points which are reasonably consistent with function y = f(x)

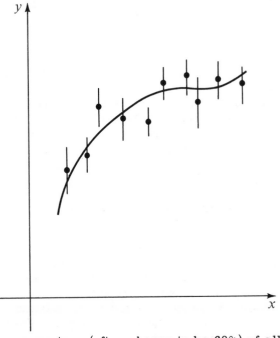

percentage (often chosen to be 68%) of all measurements fall within $\pm\Delta y_j$ of the average value $\bar{y}_j$. If the theoretical curve $y = f(x)$ is correct, as the number of repeated measurements at the same value of x_j becomes infinite, we expect $\bar{y}_j = f(x_j)$, i.e., we expect the average of all measurements to lie right on the curve. Furthermore, for any single measurement y_j, there is a 68% chance that the residual r_j is less than Δy_j. Thus, there is a 68% chance that $\left|\frac{r_j}{\Delta y_j}\right| < 1$ (error bar intersects curve), and a 32% chance that $\left|\frac{r_j}{\Delta y_j}\right| > 1$ (error bar does not intersect curve).

Chi Square and Probability

For many kinds of measurements we find that the quantity $t_j = \frac{r_j}{\Delta y_j}$ is distributed according to the well-known Gaussian distribution $\exp(-t_j{}^2)$. The quantity t_j is a measure of the discrepancy between the jth data point and the curve. To find the overall discrepancy between the entire set of data points and the curve, the so-called goodness-of-fit, we cannot simply take the sum $\sum\limits_{j=1}^{n} t_j = \sum\limits_{j=1}^{n} \frac{r_j}{\Delta y_j}$ since the residuals r_j are likely to be positive and negative about equally often, making the sum close to zero. The quantity χ^2 (chi square), defined as the sum of the squares of the t_j's:

$$\chi^2 = \sum_{j=1}^{n} t_j{}^2 = \sum_{j=1}^{n} \left(\frac{r_j}{\Delta y_j}\right)^2, \tag{5.2}$$

Table 5.1   Probability of a larger value of χ^2

n	P = .99	.98	.95	.90	.80	.70	.50	.30	.20	.10	.05	.02	.01
1	.000157	.000628	.00393	.0158	.0642	.148	.455	1.074	1.642	2.706	3.841	5.412	6.635
2	.0201	.0404	.103	.211	.446	.713	1.386	2.408	3.219	4.605	5.991	7.824	9.210
3	.115	.185	.352	.584	1.005	1.424	2.366	3.665	4.642	6.251	7.815	9.837	11.345
4	.297	.429	.711	1.064	1.649	2.195	3.357	4.878	5.989	7.779	9.488	11.668	13.277
5	.554	.752	1.145	1.610	2.343	3.000	4.351	6.064	7.289	9.236	11.070	13.388	15.086
6	.872	1.134	1.635	2.204	3.070	3.828	5.348	7.231	8.558	10.645	12.592	15.033	16.812
7	1.239	1.564	2.167	2.833	3.822	4.671	6.346	8.383	9.803	12.017	14.067	16.622	18.475
8	1.646	2.032	2.733	3.490	4.594	5.527	7.344	9.524	11.030	13.362	15.507	18.168	20.090
9	2.088	2.532	3.325	4.168	5.380	6.393	8.343	10.656	12.242	14.684	16.919	19.679	21.666
10	2.558	3.059	3.940	4.865	6.179	7.267	9.342	11.781	13.442	15.987	18.307	21.161	23.209
11	3.053	3.609	4.575	5.578	6.989	8.148	10.341	12.899	14.631	17.275	19.675	22.618	24.725
12	3.571	4.178	5.226	6.304	7.807	9.034	11.340	14.011	15.812	18.549	21.026	24.054	26.217
13	4.107	4.765	5.892	7.042	8.634	9.926	12.340	15.119	16.985	19.812	22.362	25.472	27.688
14	4.660	5.368	6.571	7.790	9.467	10.821	13.339	16.222	18.151	21.064	23.685	26.873	29.141
15	5.229	5.985	7.261	8.547	10.307	11.721	14.339	17.322	19.311	22.307	24.996	28.259	30.578
16	5.812	6.614	7.962	9.312	11.152	12.624	15.338	18.418	20.465	23.542	26.296	29.633	32.000
17	6.408	7.255	8.672	10.085	12.002	13.531	16.338	19.511	21.615	24.769	27.587	30.995	33.409
18	7.015	7.906	9.390	10.865	12.857	14.440	17.338	20.601	22.760	25.989	28.869	32.346	34.805
19	7.633	8.567	10.117	11.651	13.716	15.352	18.338	21.689	23.900	27.204	30.144	33.687	36.191
20	8.260	9.237	10.851	12.443	14.578	16.266	19.337	22.775	25.038	28.412	31.410	35.020	37.566
21	8.897	9.915	11.591	13.240	15.445	17.182	20.337	23.858	26.171	29.615	32.671	36.343	38.932
22	9.542	10.600	12.338	14.041	16.314	18.101	21.337	24.939	27.301	30.813	33.924	37.659	40.289
23	10.196	11.293	13.091	14.848	17.187	19.021	22.337	26.018	28.429	32.007	35.172	38.968	41.638
24	10.856	11.992	13.848	15.659	18.062	19.943	23.337	27.096	29.553	33.196	36.415	40.270	42.980
25	11.524	12.697	14.611	16.473	18.940	20.867	24.337	28.172	30.675	34.382	37.652	41.566	44.314
26	12.198	13.409	15.379	17.292	19.820	21.792	25.336	29.246	31.795	35.563	38.885	42.856	45.642
27	12.879	14.125	16.151	18.114	20.703	22.719	26.336	30.319	32.912	36.741	40.113	44.140	46.963
28	13.565	14.847	16.928	18.939	21.588	23.647	27.336	31.391	34.027	37.916	41.337	45.419	48.278
29	14.256	15.574	17.708	19.768	22.475	24.577	28.336	32.461	35.139	39.087	42.557	46.693	49.588
30	14.953	16.306	18.493	20.599	23.364	25.508	29.336	33.530	36.250	40.256	43.773	47.962	50.892

Table 5.1 is taken from Table III of Fisher: *Statistical Methods for Research Workers*, published by Oliver and Boyd, Edinburgh, by permission of the author and publishers.

is a measure of the goodness-of-fit of a theoretical curve $y = f(x)$ to a set of data points. For the best conceivable fit (all the data points on the curve), we find $\chi^2 = 0$; while for a reasonably good fit (all the residuals r_j comparable to the measurement errors Δy_j), we find $\chi^2 \approx n$, the number of data points.

Based on the assumed (Gaussian) distribution for each of the t_j's, it is possible to calculate the probability that χ^2 has a particular value. This is usually expressed in terms of a cumulative probability, $P(\chi^2, n)$ or *confidence level,* which is given in Table 5.1. The number $P(\chi^2, n)$ is the probability of finding a value for χ^2 at least as great as the value actually computed, *assuming* that the hypothesized function $y = f(x)$ is correct and that the deviations for each data point are *only* the result of random measurement error.

The first column in the table lists n values from 1 to 30. The table entries in the row for each n value are the values of χ^2 corresponding to the confidence level listed as column headings. As can be seen in Table 5.1, for a given value of n, a value of χ^2 which greatly exceeds n is associated with a low confidence level. A low confidence level means that the fit is a poor one, and that it is unlikely that the hypothesized relation $y = f(x)$ is consistent with the data. To illustrate these points, suppose we do not know which of three hypothesized functions $f_1(x), f_2(x)$, or $f_3(x)$, describes the relation between the variables y and x. (See Figure 5.2). For the three functions we use equations 5.1 and 5.2 to compute three values of χ^2: $\chi_1^2, \chi_2^2, \chi_3^2$, which indicate the goodness-of-fit for the three functions. We can use these three values in a *chi square test* in an attempt to determine which of the three functions is correct. Let us assume that the three computed values are $\chi_1^2 = 6.2, \chi_2^2 = 10.8, \chi_3^2 = 23.2$. We then use the chi square table (with n = 10) to find the three probabilities corresponding to these values of χ^2: $P_1 = .80, P_2 = .40, P_3 = .01$. (An interpolation between table entries is necessary to get P_2.) The result $P_3 = .01$ means that only 1% of the time would *random* errors result in so large a value of χ^2, assuming that $y = f_3(x)$ is the correct relation between y and x. *We normally regard a confidence level this low as grounds for **rejecting** the hypothesis that $f_3(x)$ is the correct relation between y and x.* For the functions $f_1(x)$ and $f_2(x)$, the respective probabilities 80% and 40% mean that $f_1(x)$ is a somewhat better fit than $f_2(x)$. However, we do not reject $f_2(x)$ simply because the probability P_2 is only half P_1; a 40% confidence level is *not* so low that the hypothesis can be rejected.[*] Furthermore, we cannot even claim that the probability of $f_1(x)$ being the correct function is twice as great as the probability of $f_2(x)$ being the correct function. The only conclusion we can draw is that from the chi square test we cannot decide which of the two functions $f_1(x)$ or $f_2(x)$ is the correct one, since both give reasonably good fits to the data. From this example, we see that the chi square test is useful in *rejecting* hypotheses, but it cannot be used to prove the correctness of a particular hypothesis (unless every other possible hypothesis is rejected).

Apart from finding the goodness of fit for comparing specific functions (such as $f_1(x), f_2(x)$, and $f_3(x)$), the chi square test can find the goodness-of-fit for a

---

[*] This rather subtle point can be understood by recalling the meaning of the probabilities P_1 and P_2: P_1 is the probability of obtaining a chi square as large as the value actually found *assuming* $f_1(x)$ is the correct function; P_2 is the probability of obtaining a chi square as large as that actually found *assuming* $f_2(x)$ is the correct function. Clearly, the ratio P_1/P_2 has no bearing on the relative probabilities of $f_1(x)$ and $f_2(x)$ actually being the correct functions.

Figure 5.2   A set of data points and three hypothesized functions

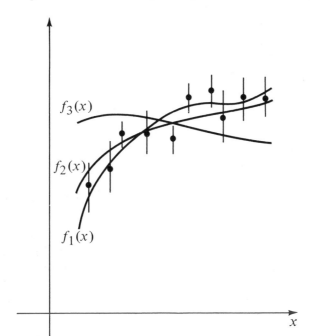

function f(x) which is not completely specified, but which has a number of "adjustable" parameters $\alpha_1, \alpha_2, \ldots, \alpha_m$, whose values are not known a priori. In fact, one reason for the chi square test is to determine the values of the parameters which give the best fit to the data. In other words, we wish to determine the values of the parameters $\alpha_1, \alpha_2, \ldots, \alpha_m$, which give the smallest value for chi square. This is known as the method of *least squares* since it involves minimizing the sum of the squares of deviations (equation 5.2) with respect to the adjustable parameters $\alpha_1, \alpha_2, \ldots, \alpha_m$. The function f(x) which has the minimum chi square is called the *least squares fit,* and according to the chi square criterion it is the "best" fit to the data. At the same time we find the least squares fit, we can also determine if the fit is, in fact, a good one: Does the value of χ^2 correspond to a reasonably high confidence level? In using the chi square table to find the confidence level for a fit to n data points which involves m adjustable parameters, we must use the difference n − m (rather than n itself). The difference n − m is known as the number of *degrees of freedom.*

One-Parameter Least Squares Fit

The simplest type of least squares fit involves a single adjustable parameter α, which multiplies a function f(x) that is otherwise completely specified: $y = \alpha\, f(x)$. In this case, we may write for χ^2:

$$\chi^2 = \sum_j \frac{r_j^2}{\Delta y_j^2} = \sum_j \frac{(y_j - \alpha f(x_j))^2}{\Delta y_j^2} \tag{5.3}$$

To find the value of the parameter α which minimizes χ^2, we set the partial derivative $\frac{\partial \chi^2}{\partial \alpha} = 0$:

$$\frac{\partial \chi^2}{\partial \alpha} = \sum_j \frac{-2f(x_j)(y_j - \alpha f(x_j))}{\Delta y_j^2} = 0$$

Solving for α, we obtain with the aid of a little algebra:

$$\alpha^* = \left(\sum_j \frac{f(x_j)\, y_j}{\Delta y_j^2}\right) \bigg/ \left(\sum_j \frac{(f(x_j))^2}{\Delta y_j^2}\right), \tag{5.4}$$

where the asterisk indicates that this is the value of α which gives the minimum value of χ^2.

We can obtain an estimate of the uncertainty in α by determining the change in α from the least-squares fit value α^* which decreases the goodness-of-fit (i.e., increases χ^2) by some specified amount. The uncertainty $\Delta\alpha$ is usually defined so that the value of χ^2 for both $\alpha = \alpha^* + \Delta\alpha$ and $\alpha = \alpha^* - \Delta\alpha$ is equal to $\chi^2_{min} + 1$, where $\chi^2 = \chi^2_{min}$ for $\alpha = \alpha^*$. (See Figure 5.3.) It can be shown that for a one-parameter fit, the uncertainty $\Delta\alpha$ is given by

$$\Delta\alpha = \left(\sum_j \frac{(f(x_j))^2}{\Delta y_j^2}\right)^{-1/2}. \tag{5.5}$$

Figure 5.3   A plot of χ^2 against α

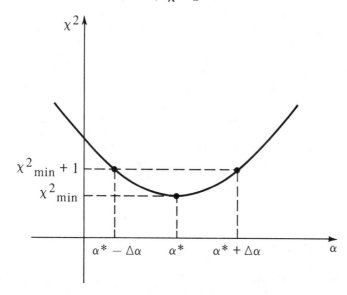

A special case of a one-parameter fit that occurs quite frequently is $f(x) = 1$, which gives $y = \alpha$. In this case, the least squares fit value of α^* given by equation 5.4 is the y-intercept of the horizontal straight line which gives the best fit to the data points $(x_1, y_1), \ldots, (x_n, y_n)$. In effect, α^* is the weighted average of $y_1, \ldots, y_n$, with weights inversely proportional to the square of the measurement errors $\Delta y_1, \ldots, \Delta y_n$.

As an example of this special case, we list below nine independent measurements $(y_1, \ldots, y_9)$ of the mass of the omega meson (a subatomic particle), together with estimated measurement errors $(\Delta y_1, \ldots, \Delta y_9)$ from nine different experiments.

Mass (in units of MeV/c^2)	Error	Year reported
779.4	1.4	1962
784.0	1.0	1963
781.0	2.0	1964
785.6	1.2	1965
786.0	1.0	1966
779.5	1.5	1967
784.8	1.1	1968
786.0	1.0	1969
784.1	1.2	1970

For the nine x-values $x_1, \ldots, x_9$, we use the year each measurement was reported. Since the mass of a subatomic particle does not depend on the year the measurement was made, the curve we wish to fit to the nine data points, shown plotted in Figure 5.4, is a horizontal line $y = \alpha$. To find the value of α which minimizes χ^2, we use equation 5.4 with $f(x) = 1$:

$$\alpha^* = \left(\sum_j \frac{y_j}{\Delta y_j^2} \right) \Big/ \left(\sum_j \frac{1}{\Delta y_j^2} \right).$$

Upon substitution of the nine masses $(y_1, \ldots, y_9)$ and measurement errors $(\Delta y_1, \ldots, \Delta y_9)$, we find $\alpha^* = 784.12$. Similarly, to find the estimated uncertainty $\Delta \alpha$, we use equation 5.5 with $f(x) = 1$:

$$\Delta \alpha = \left(\sum_j \frac{1}{\Delta y_j^2} \right)^{-1/2},$$

from which we find $\Delta \alpha = 0.39$. Thus the value $\alpha = 784.12 \pm 0.39$ represents the weighted average of the omega meson's mass found from a one-parameter least squares fit to the nine experimental values. The solid horizontal line in Figure 5.4 indicates the value of α^* and the dotted horizontal lines indicate the band corresponding to the range $\alpha^* \pm \Delta \alpha$.

In addition to finding the value of one or more parameters from a least squares fit, we can use the chi square test to determine if this best fit is any good. By substitution of $\alpha = 784.12$, $f(x) = 1$, and the experimental values for $y_1, \ldots, y_9$; $\Delta y_1, \ldots, \Delta y_9$, in equation 5.3, we obtain the minimum value of χ^2: $\chi^2_{min} = 32.3$. According to the chi square table, the probability of finding a chi square this high for eight degrees of freedom is less than .01. In other words, the "best" (i.e., minimum χ^2) fit is no good in this case. This actually should be obvious from a visual inspection of the data points in Figure 5.4, since there is no horizontal line which would pass reasonably close to all the data points.

If all the data points and measurement errors are correct, the chi square test gives grounds for rejecting the hypothesis that the mass of the omega meson is a

Figure 5.4   Plot of ω^0 mass against year reported

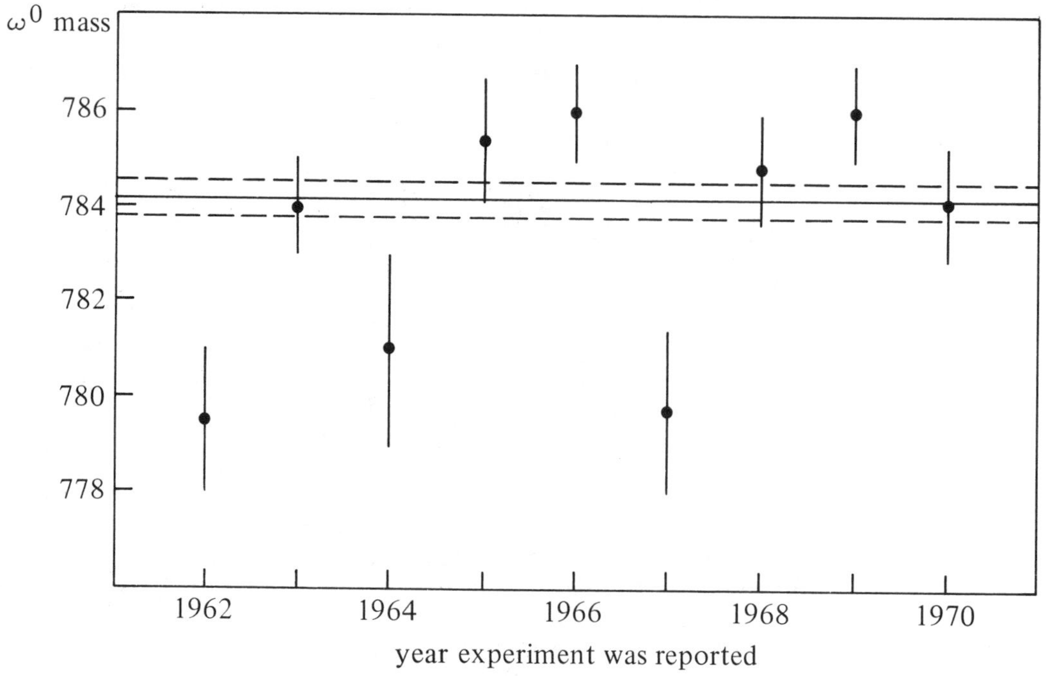

year experiment was reported

constant, independent of the year it is measured! A more plausible interpreta-tion of the poor fit is that one or more of the reported measurements is in error. We notice, for example, that if the measurements made in 1962, 1964, and 1967 are dropped, the remaining six measurements are quite consistent with a hori-zontal line. However, arbitrarily dropping these three data points in order to improve the fit would probably be unwise, because it seems likely that some of the experimenters made *systematic* (i.e., nonrandom) errors, and statistical tests cannot be used to prove which experiments are in error. In practice, of course, if one experiment out of a large number gives a significantly different result from the others, we might be tempted to discard that result. However, this should not be done unless we can be reasonably certain that the anomalous result cannot be ascribed to better apparatus or other favorable conditions which were absent in all the other experiments, causing *them* to be in error.

Two-Parameter Least Squares Fit

A two-parameter least squares fit to a set of data can be easily made if the relation between y and x is *linear* in the parameters α_1 and α_2:

$$y = \alpha_1 f_1(x) + \alpha_2 f_2(x),$$

where $f_1(x)$ and $f_2(x)$ are two specified functions. We shall consider the im-portant special case: $f_1(x) = 1$ and $f_2(x) = x$, so that we wish to make a least squares fit to the straight line $y = \alpha_1 + \alpha_2 x$. In order to minimize χ^2 with respect to the two parameters α_1 and α_2, we require that the two partial derivatives $\dfrac{\partial \chi^2}{\partial \alpha_1}$ and $\dfrac{\partial \chi^2}{\partial \alpha_2}$ vanish. Thus, with χ^2 given by

$$\chi^2 = \sum_j \frac{(y_j - \alpha_1 - \alpha_2 x_j)^2}{\Delta y_j^2},$$

we require that

$$\frac{\partial \chi^2}{\partial \alpha_1} = \sum_j \frac{-2(y_j - \alpha_1 - \alpha_2 x_j)}{\Delta y_j^2} = 0 \tag{5.6}$$

and

$$\frac{\partial \chi^2}{\partial \alpha_2} = \sum_j \frac{-2x_j(y_j - \alpha_1 - \alpha_2 x_j)}{\Delta y_j^2} = 0. \tag{5.7}$$

Solving equations 5.6 and 5.7 simultaneously for the two unknowns α_1 and α_2, we obtain the values of α_1 and α_2 which minimize χ^2:

$$\alpha_1^* = \frac{1}{\Delta}(S_{xx}S_y - S_x S_{xy}) \tag{5.8}$$

$$\alpha_2^* = \frac{1}{\Delta}(S_1 S_{xy} - S_x S_y),$$

where we have used the following abbreviations:

$$S_1 = \sum_j \frac{1}{\Delta y_j^2} \qquad S_x = \sum_j \frac{x_j}{\Delta y_j^2} \qquad S_y = \sum_j \frac{y_j}{\Delta y_j^2}$$

$$\Delta = S_1 S_{xx} - S_x^2 \qquad S_{xx} = \sum_j \frac{x_j^2}{\Delta y_j^2} \qquad S_{xy} = \sum_j \frac{x_j y_j}{\Delta y_j^2}$$

As in the case of the one-parameter least squares fit, each of the parameters α_1 and α_2 has an uncertainty, which here indicates the amount each of the parameters must be *independently*[*] changed from the least squares fit values α_1^* and α_2^*, in order to increase the computed value of χ^2 to one greater than the minimum value $\chi^2_{\min}$. It can be shown that the two uncertainties $\Delta\alpha_1$ and $\Delta\alpha_2$ are given by

$$\Delta\alpha_1 = \left(\frac{S_{xx}}{\Delta}\right)^{1/2}$$

$$\Delta\alpha_2 = \left(\frac{S_1}{\Delta}\right)^{1/2} \tag{5.9}$$

---

[*] In least squares fits involving two or more adjustable parameters, we can define "correlation terms," which indicate how the value of χ^2 changes when the parameters are simultaneously varied. In the case of the two-parameter fit, there is one correlation term: $\Delta\alpha_{12} = -\dfrac{S_x}{\Delta}$.

Least squares fits can, of course, be found for functions which are not linear, but the procedure is more complex. Therefore, it is usually desirable to try to transform the measured variables to other variables which do obey a linear relation. For example, if we measure the voltage V across the plates of a discharging capacitor at time t we expect the relation between V and t to be

$$V = V_0 e^{-t/T} \tag{5.10}$$

where V_0 is the voltage at time $t = 0$ and T is the "time constant" (see page 76). To obtain variables that are linearly related, we take the natural logarithm of both sides of equation 5.10, to obtain

$$\ln V = \ln V_0 - t/T. \tag{5.11}$$

Thus, the transformations $\ln V \rightarrow y, t \rightarrow x$ give a linear relation between x and y in equation 5.11.

Another pertinent reason for transforming the measured quantities to obtain a linear relation is that we can then often tell by visual inspection whether a straight line is a good fit to the plotted data points, thereby testing the original hypothesized relation. As an illustration, we consider the following set of data obtained from a simple experiment using a cart sliding down an incline with negligible friction (an air track), inclined at an angle $\theta = .005546$ radians. (One end of the air track rests on a half-inch metal block.) The data represent the measured times for a cart released from rest to travel various distances along the track. The distances $d_1, d_2, \ldots, d_{15}$, are in centimeters, and the times $t_1, t_2, \ldots, t_{15}$, are in seconds.

t	0.7	1.3	1.9	2.5	3.1	3.7	4.1	4.9	5.6	6.1	6.7	7.5	7.9	8.5	9.1
d	1	4	9	16	25	36	49	64	81	100	121	144	169	196	225

We expect the data to be consistent with the relation:

$$d = \tfrac{1}{2} g \sin \theta \, t^2, \tag{5.12}$$

where g is the acceleration due to gravity. We can solve for t to obtain

$$t = \left(\frac{2}{g \sin \theta}\right)^{1/2} d^{1/2}. \tag{5.13}$$

We can therefore get a linear relation if we make the transformation $t \rightarrow y$, $d^{1/2} \rightarrow x$. The transformed data points (t against $d^{1/2}$) are plotted in Figure 5.5. The small error bars on each point indicate the measurement errors in t. Based on repeated measurements for each point, the error was estimated to be ±0.1 seconds for all points. The data points seem to be reasonably consistent with a straight line $y = \alpha_1 + \alpha_2 x$, where $\alpha_1 = 0$. In fact, a fairly good straight line fit to the data can be found by simply drawing a straight line with the data points distributed on either side of it in a random manner.

For a more precise result, we can find the slope α_2, y-intercept α_1, and the associated errors $\Delta\alpha_2$ and $\Delta\alpha_1$, for the least squares fit straight line, using equations 5.8 and 5.9. We can determine a value for g, the acceleration due to

Figure 5.5   Plot of $d^{1/2}$ against t for data from air-track experiment

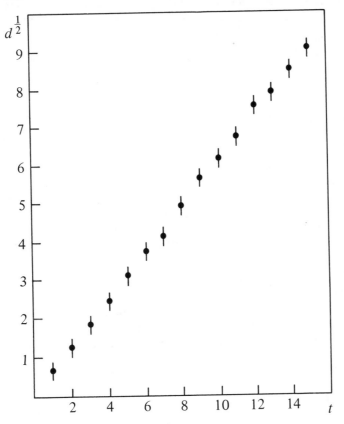

gravity, from the slope of the line. According to the form of equation 5.13, the slope α_2 is given by

$$\alpha_2 = \left(\frac{2}{g \sin \theta}\right)^{1/2},$$ (5.14)

which can be solved for g, to obtain

$$g = \frac{2}{\alpha_2^2 \sin \theta} .$$ (5.15)

The theory of first-order error propagation can be used to find the uncertainty in a quantity calculated from a known function of the parameters α_1 and α_2. Since g depends only on α_2, we can find the magnitude of the error in g using

$$\Delta g = \left|\frac{dg}{d\alpha_2}\right| \Delta\alpha_2.$$ (5.16)

From equations 5.15 and 5.16, we find that the percentage error in g is twice the percentage error in α_2:

$$\frac{\Delta g}{g} = 2\,\frac{\Delta\alpha_2}{\alpha_2} .$$ (5.17)

In this example, we could have used our a priori knowledge that the y-intercept should be zero, and found a one-parameter least squares fit to the function $y = \alpha x$. We would use equations 5.4 and 5.5 with $f(x) = x$ to obtain

$$\alpha^* = \left(\sum_j \frac{x_j y_j}{\Delta y_j^2}\right) \bigg/ \left(\sum_j \frac{x_j^2}{\Delta y_j^2}\right)$$

and

$$\Delta\alpha = \left(\sum_j \frac{x_j^2}{\Delta y_j^2}\right)^{-1/2}.$$

One advantage of the one-parameter fit is that a smaller value for the uncertainty in the slope $\Delta\alpha$ would probably be found due to the more constrained nature of the fit. On the other hand, the two-parameter fit has the advantage that we may possibly establish the existence of a systematic error in the data should we find that α_1^* (the y-intercept) is significantly different from zero.

Least Squares Fit with Unknown Measurement Errors

We can find a least squares fit even in cases where the measurement errors $\Delta y_1, \ldots, \Delta y_n$, are unknown, by assuming that the errors are the same for all points. If we call the common (but unknown) error Δy, then equation 5.2 becomes

$$\chi^2 = \frac{1}{\Delta y^2} \sum_{j=1}^{n} r_j^2. \tag{5.18}$$

Since Δy is a constant, we can minimize χ^2 with respect to the parameters α_1, $\alpha_2, \ldots, \alpha_m$, without knowing the value of Δy. However, if we wish to determine the uncertainties in the parameters for the best fit: $\Delta\alpha_1, \Delta\alpha_2, \ldots, \Delta\alpha_m$, a value for Δy *is* needed. A reasonable estimate for Δy is the root mean square value of the residuals:

$$\Delta y = \sqrt{\frac{\sum r_j^2}{n}}, \tag{5.19}$$

which is a measure of the average "scatter" of the points about the least squares fit curve.

Thus it is possible to find values for the parameters $\alpha_1, \alpha_2, \ldots, \alpha_m$, and estimated uncertainties $\Delta\alpha_1, \Delta\alpha_2, \ldots, \Delta\alpha_m$, for the least squares fit, even though the measurement errors $\Delta y_1, \Delta y_2, \ldots, \Delta y_n$ are unknown. However, it is *not* possible to use the computed value of χ^2 to tell if the "best" fit is any good—because by assuming the value for Δy, we have, in effect, assumed a value for χ^2 equal to n, as can be seen by substitution of equation 5.19 into 5.18. In such a case, the only way to tell if the least squares fit is any good is to see if the data points are distributed about the curve in an apparently random manner, that is, to check that the residuals r_j lack a systematic variation from point to point. For example, consider the two cases shown in Figures 5.6a and 5.6b. Even though many of the data points in (a) are closer to the curve than those in (b), we would tend to judge (b) as the better fit. We would suspect that (a) is not a good fit even if we were told that every point is actually closer to the curve than the estimated measurement error Δy. Experimental data are very unlikely to deviate from a curve in such a systematic way (although it is mathematically possible). The most likely possibility is that the experimenter has overestimated his measurement error, and the data points are not consistent with the curve. There is, of course, no way to infer from this whether it is the experiment or the theory (or possibly both) that is incorrect.

Figure 5.6   A comparison of two fits

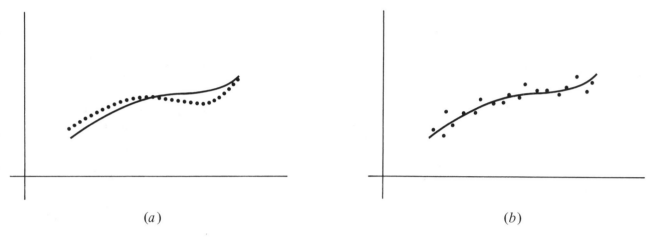

(*a*)   (*b*)

Least Squares Straight Line Fitting Program

The program given in flow diagram form in Figures 5.7 through 5.9, and listed on pages 238-241, does a two-parameter least squares fit to a straight line, given a set of n data points $(x_1, y_1), \ldots, (x_n, y_n)$. A set of n measurement errors $\Delta y_1, \ldots, \Delta y_n$ are optional. In addition to computing values for $\alpha_1, \alpha_2, \Delta\alpha_1,$ $\Delta\alpha_2,$ and χ^2, the program also makes a plot of the least squares fit straight line, together with the data points.

The program reads a value for N, the number of data points (n), and the first of N more data cards, each containing a data point and its y-error: $x_j, y_j, \Delta y_j$. If the measurement errors are unknown, the value of Δy_j should be zero on all cards. The program then goes through a loop in which it reads the remaining N-1 data cards, computes the various sums needed to calculate α_1 and α_2, and finds the largest and smallest values of x_j and y_j (needed for plotting purposes). Note that if Δy_j is zero on a data card, indicating an unknown measurement error, the value actually used in computing the sums $S_1, S_x, \ldots, S_{xy}$, is 1.0. The program calculates the parameters α_1 and α_2 for the least squares fit straight line, using equations 5.8. It then enters a loop in which it computes χ^2 and the residuals for each data point, and prints these together with $x_j, y_j,$ and Δy_j. The program then computes the errors in the parameters $\Delta\alpha_1$ and $\Delta\alpha_2$, according to equations 5.9. Note that if the value of Δy_1 is zero, then it is assumed that the measurement errors are unknown, and the value used for the error Δy (the same for all points) is the root mean of the residuals (equation 5.19). The uncertainties $\Delta\alpha_1$ and $\Delta\alpha_2$ are then recalculated accordingly. After it prints the values of $\alpha_1, \alpha_2, \Delta\alpha_1, \Delta\alpha_2,$ and χ^2, the program enters a final loop in which it

Figure 5.7   Flow diagram for least squares fitting program—Part 1

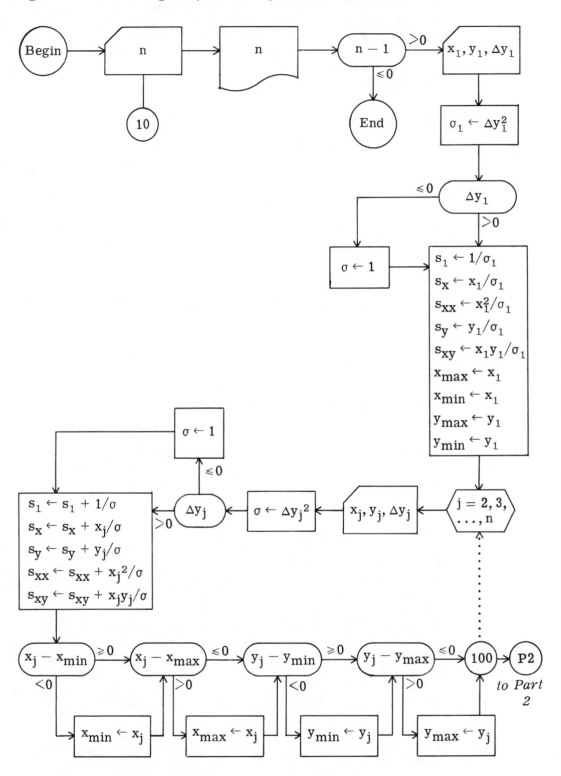

Figure 5.8   Flow diagram for least squares fitting program—Part 2

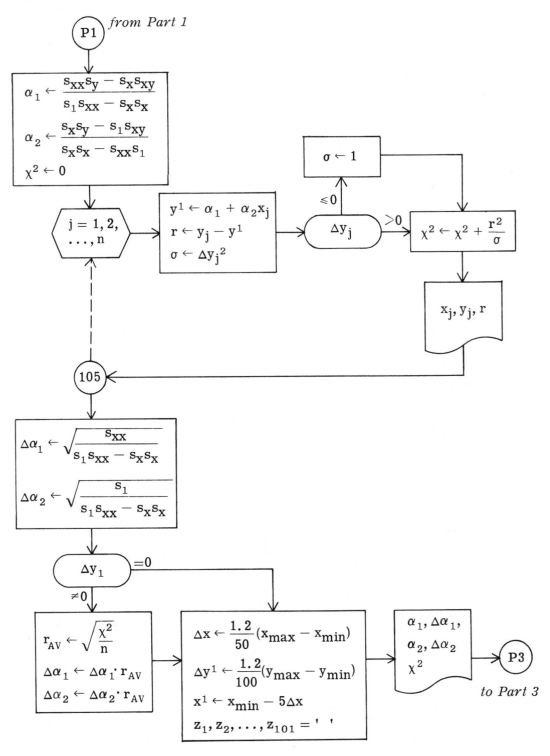

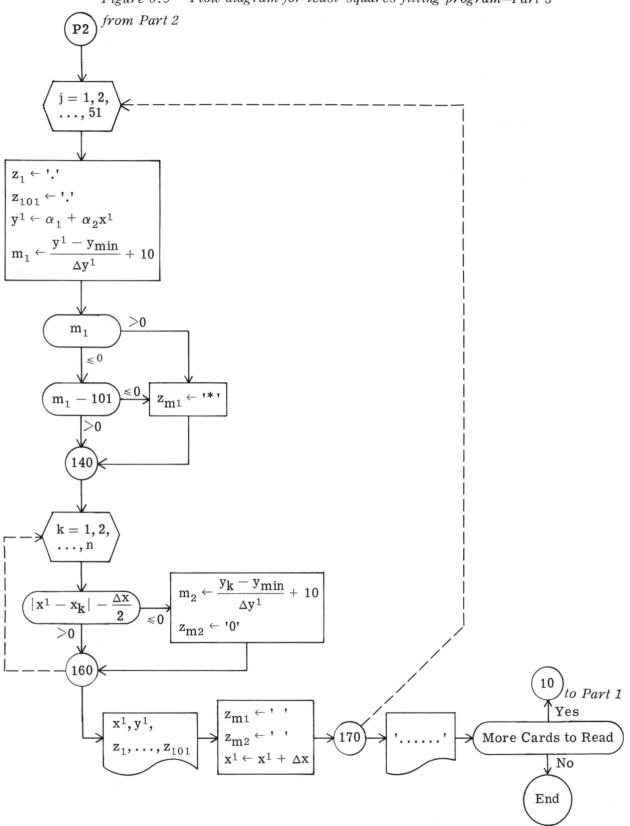

Figure 5.9   Flow diagram for least-squares fitting program—Part 3

gives both a numerical table of x and y values and a plot. The least squares fit straight line is plotted with asterisks and the data points are shown by the letter O (see Figure 5.11). Note that the value printed for χ^2 is not meaningful in the case of unknown measurement errors (Δy_j set to zero on the data cards).

Comments on Data Cards and Sample Results

The output shown in Figures 5.10-5.11 was obtained using N = 15 and 15 data cards with the values of x_j, y_j, Δy_j which are printed in Figure 5.10. This is the set of data from the air track experiment discussed on page 230. The number on the first data card (15.0) is the number of data points. For the remaining cards, the first number is the value of the x variable ($d^{1/2}$), the second number is the value of the y variable (t), and the third number is the value of the measurement error in y, which is estimated as 0.1 seconds for all points.

Figure 5.10

NUMBER OF DATA POINTS = 15.00000

X	Y	DY	RESIDUALS
1.00000	.70000	.10000	.02083
2.00000	1.30000	.10000	.01690
3.00000	1.90000	.10000	.01298
4.00000	2.50000	.10000	.00905
5.00000	3.10000	.10000	.00512
6.00000	3.70000	.10000	.00119
7.00000	4.10000	.10000	-.20274
8.00000	4.90000	.10000	-.00667
9.00000	5.60000	.10000	.08940
10.00000	6.10000	.10000	-.01452
11.00000	6.70000	.10000	-.01845
12.00000	7.50000	.10000	.17762
13.00000	7.90000	.10000	-.02631
14.00000	8.50000	.10000	-.03024
15.00000	9.10000	.10000	-.03417

The output in Figure 5.10 lists the computed residuals, $r_j = y_j - f(x_j)$, for each data point. The residuals are all reasonably small compared to the measurement errors; in only one case does a residual exceed twice the error. There may, however, be some systematic variation in the residuals, which would indicate the presence of a small systematic error. The least squares fit straight line plotted in Figure 5.11 appears to be quite consistent with the data points shown as circles. In order to use the χ^2 table on page 223 to determine the confidence level for the fit, we need to specify the number of degrees of freedom: 13 (the number of data points (15) minus the number of adjustable parameters (2)). The value of χ^2 for the least squares fit is 8.5 in the output in Figure 5.11. According to the χ^2 table, the confidence level for a χ^2 of 8.5 with 13 degrees of freedom is about 0.80, or 80 percent, which means the fit is very good. The values of the parameters α_1, α_2, and their errors, for the least squares fit straight line are given in Figure 5.11:

$$\alpha_1 = 0.07523 \pm 0.05433$$
$$\alpha_2 = 0.60392 \pm 0.00597$$

Figure 5.11

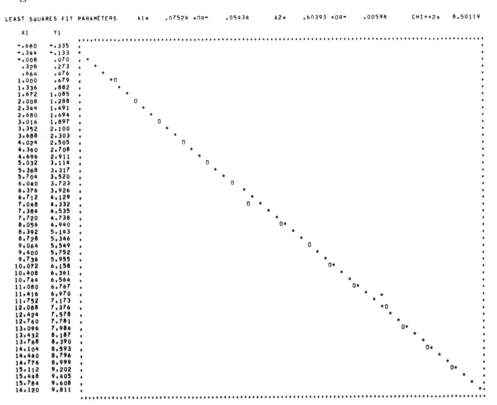

The value of the y-intercept α_1 is almost consistent with zero. The fact that the value is a bit further from zero than the uncertainty $\Delta\alpha_1$ is probably not significant.* From the value for the slope α_2, we can compute values for g and Δg using equations 5.15 and 5.17:

$$g = 986 \pm 20 \text{ cm/sec}^2,$$

which is quite consistent with the accepted value, 980 cm/sec^2.

Problems 1-5 at the end of the chapter pertain to the least squares fitting of experimental data.

```
C          LEAST SQUARES STRAIGHT LINE FIT
C
C--------------------------------------------------------------------------------
C
C          THIS PROGRAM FINDS A TWO-PARAMETER FIT TO A STRAIGHT LINE.
C
```

---

*   The uncertainties calculated in a least squares fit correspond to one standard deviation limits on the parameters.

```
       REAL N
       DIMENSION X(100),Y(100),DY(100),Z(101)
       DATA BLANK,DOT,ASTER,OH/' ','.','*','O'/
C
C              READ N, THE NUMBER OF DATA POINTS, FROM THE FIRST CARD.
C
   10  READ(2,1000)N
       WRITE(3,1005)N
       IF(N-1.0)180,180,20
C
C              READ COORDINATES X(1),Y(1), AND ERROR DY(1) FOR FIRST POINT.
C
   20  READ(2,1001)X(1),Y(1),DY(1)
C
C              STORE INITIAL VALUES IN SUMS.
C
       SIG=DY(1)**2
       IF(DY(1))22,22,24
   22  SIG=1.0
   24  CONTINUE
       S1=1.0/SIG
       SX=X(1)/SIG
       SXX=X(1)**2/SIG
       SY=Y(1)/SIG
       SXY=X(1)*Y(1)/SIG
       XMAX=X(1)
       XMIN=X(1)
       YMIN=Y(1)
       YMAX=Y(1)
       M=N
       DO 100 J=2,M
C
C              READ A DATA CARD CONTAINING X(J),Y(J),DY(J).
C
       READ(2,1001)X(J),Y(J),DY(J)
C
C              COMPUTE SUMS.
C
       SIG=DY(J)**2
       IF(DY(J))26,26,28
   26  SIG=1.0
   28  CONTINUE
       S1=S1+1.0/SIG
       SX=SX+X(J)/SIG
       SY=SY+Y(J)/SIG
       SXX=SXX+X(J)**2/SIG
       SXY=SXY+X(J)*Y(J)/SIG
C
C              FIND MAXIMUM AND MINIMUM VALUES OF X AND Y.
C
       IF(X(J)-XMIN)30,40,40
   30  XMIN=X(J)
   40  IF(X(J)-XMAX)60,60,50
   50  XMAX=X(J)
   60  IF(Y(J)-YMIN)70,80,80
   70  YMIN=Y(J)
   80  IF(Y(J)-YMAX)100,100,90
   90  YMAX=Y(J)
  100  CONTINUE
C
C              COMPUTE  PARAMETERS A1 AND A2.
C
```

```
      A1=(SXX*SY-SX*SXY)/(S1*SXX-SX*SX)
      A2=(SX*SY-S1*SXY)/(SX*SX-SXX*S1)
C
C             COMPUTE RESIDUALS R  AND CHI SQUARE (CHI).
C
      CHI=0.
      DO 105 J=1,M
      Y1=A1+A2*X(J)
      R=Y(J)-Y1
      SIG=DY(J)**2
      IF(DY(J))101,101,102
  101 SIG=1.0
  102   CHI=CHI+R**2/SIG
      WRITE(3,1006)X(J),Y(J),DY(J),R
  105 CONTINUE
C
C             COMPUTE UNCERTAINTIES DA1 AND DA2.
C
      DA1=SQRT(SXX/(S1*SXX-SX*SX))
      DA2=SQRT(S1/(S1*SXX-SX*SX))
C
C             IF DY(1) IS ZERO, THEN ASSUME ERRORS ARE UNKNOWN.
C
      IF(DY(1))109,107,109
  107 RAV=SQRT(CHI)/N
      DA1=DA1*RAV
      DA2=DA2*RAV
  109 CONTINUE
C
C             DX AND DY1 ARE STEP SIZES (USED FOR MAKING PLOT).
C
      DX=1.2*(XMAX-XMIN)/50.
      DY1=1.2*(YMAX-YMIN)/100.
C
C             SET FIRST X VALUE FOR PLOT.
C
      X1=XMIN-5.0*DX
C
C             SET Z(1),...., Z(101) TO BLANKS.
C
      DO 110 J=1,101
  110 Z(J)=BLANK
      WRITE(3,1002)A1,DA1,A2,DA2,CHI
      DO 170 J=1,51
      Z(1)=DOT
      Z(101)=DOT
C
C             Y1 IS THE FITTED Y VALUE.
C
      Y1=A1+A2*X1
      M1=(Y1-YMIN)/DY1+10.
C
C             BE SURE THAT M1 IS BETWEEN 1 AND 101.
C
      IF(M1)140,140,120
  120 IF(M1-101)130,130,140
C
C             STORE ASTERISK IN Z(M1) FOR FITTED POINT.
C
  130 Z(M1)=ASTER
  140 CONTINUE
      DO 160 K=1,M
C
```

```
C              IF THIS X1 IS WITHIN DX OF ANY X(K), K=1,...., N,
C              THEN STORE THE LETTER O IN Z(M2).
C
       IF(ABS(X1-X(K))-DX/2.)150,150,160
   150 M2=(Y(K)-YMIN)/DY1+10.
       Z(M2)=OH
   160 CONTINUE
       WRITE(3,1003)X1,Y1,(Z(K),K=1,101)
C
C              RESTORE Z(M1) AND Z(M2) TO BLANKS.
C
       Z(M1)=BLANK
       Z(M2)=BLANK
C
C              DO NEXT FITTED X VALUE.
C
       X1=X1+DX
   170 CONTINUE
       WRITE(3,1004)
C
C              GO READ NEXT SET OF DATA, IF ANY.
C
       GO TO 10
  1000 FORMAT(F10.5)
  1001 FORMAT(3F10.5)
  1002 FORMAT(37H1LEAST SQUARES FIT PARAMETERS      A1=,F10.5,5H +OR-,
      1 F10.5, 9H     A2=,F10.5,5H +OR-,F10.5,13H      CHI**2=,F10.5//
      2 19H        X1      Y1      /,19X,101H..........................................
      3............................................................................
      4..)
  1003 FORMAT(1X,2F8.3,2X,101A1)
  1004 FORMAT(19X,101H..............................................................
      1...................................................................)
  1005 FORMAT(24H1NUMBER OF DATA POINTS =,F10.5//65H            X
      1  Y              DY              RESIDUALS/)
  1006 FORMAT(1X,4F15.5)
   180 CONTINUE
       CALL EXIT
       END
```

5.2   Analysis of Measurements from Bubble Chamber Photographs

A convenient way to study the properties of the fundamental subatomic particles is through observation of their bubble trails, or tracks, in a bubble chamber. Using measurements made directly on a bubble chamber photograph (for example, Figure 5.12), we can often identify the particles from their tracks, and calculate their masses and other properties. In a typical experiment, a beam of a particular type of particle is sent from an accelerator into a bubble chamber, which is a large liquid-filled vessel. To simplify the analysis of the data, the liquid used is often hydrogen, the simplest element. The use of liquid hydrogen, while it simplifies the analysis, complicates the experiment itself, since hydrogen, a gas at room temperature, liquifies only when cooled to −246°C. For charged

Figure 5.12

particles to leave tracks in passing through the chamber, the liquid must be in a "super-heated" state, in which the slightest disturbance causes boiling to occur. In practice, this is accomplished by expanding the vapor above the liquid with a piston a few thousandths of a second before the particles enter the chamber.

Identification of Elementary Particle Reactions

From the track that a particle leaves in the chamber, we can determine both its momentum and the sign of its charge. Since the chamber is placed in a magnetic field, the paths of charged particles are curved. A particle of charge q and momentum p which moves perpendicular to a uniform magnetic field of magnitude B travels in a circle of radius R, given by

$$R = \frac{p}{qB}.$$

For all known charged long-lived subatomic particles, the magnitude of their charge q is the same as that of the electron, $e = 1.6 \times 10^{-19}$ coulombs. Therefore, if the magnitude of the magnetic field B is known, we can determine the momentum of a particle by measuring the radius of curvature of its track. We can also infer the sign of its charge if we know the direction of the magnetic field and we observe in which sense the track curves (clockwise or counterclockwise). In the bubble chamber photographs we shall examine, the magnetic field points out of the paper, so that the sense of curvature is clockwise for positive particles and counterclockwise for negative particles, as indicated in Figure 5.13. Note that in order to determine the sign of a particle's charge, it is necessary to know in which direction the particle traveled along the track, and this is not always obvious from looking at a bubble chamber photograph.

Figure 5.13
Tracks for positive and negative particles, assuming a uniform magnetic field points out of the paper

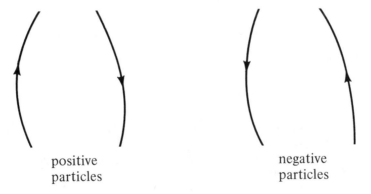

positive
particles

negative
particles

In Figure 5.12 the particles enter the chamber from the bottom of the photograph and curve counterclockwise, so they must be negatively charged. In this experiment the incoming particles are negative K mesons, which we designate K⁻, following the usual superscript notation for particles which occur in several charge states. If we follow the path of each K⁻ track, we see that some K⁻ particles passed through the chamber without incident; others interacted with a hydrogen nucleus, a proton p, and initiated various reactions. The simplest type of interaction between two particles is an elastic scattering, analogous to a billiard ball collision. Symbolically, the elastic scattering of a K⁻ particle and a proton may be written

$$K^- + p \rightarrow K^- + p.$$

An incident K⁻ meson can also initiate reactions in which new particles are created. In many cases, the created particles live for only 10^{-10} seconds or less, and then disintegrate, or *decay*, into other particles which themselves may be unstable. We shall be particularly interested in the *sigma production reaction*:

$$K^- + p \rightarrow \Sigma^- + \pi^+$$

in which the K⁻ and proton disappear and a sigma minus particle (Σ^-) and a pi plus meson (π^+) are created. In this experiment, the incident K⁻ mesons have sufficiently low energy that many of them slow down and come to rest within the chamber. In order for the sigma production reaction to take place, the K⁻ meson and the proton must be very close to one another (about 10^{-15} meters apart). Due to the attractive force between the negatively charged K⁻ and the positively charged proton, this is much more likely to happen when the K⁻ is moving slowly or at rest. There is a very simple way to determine from a bubble chamber photograph if the sigma production reaction took place when the K⁻ came to rest: since the initial momentum of the two-particle (K⁻-proton) system is zero, the Law of Conservation of Momentum requires the Σ^- and π^+ particles to have equal and opposite momentum, which makes the Σ^- and π^+ tracks appear as one continuous track. The considerable energy of the Σ^- and π^+ particles in the energy-liberating sigma production reaction is the result of the conversion of mass into energy. (The combined rest mass of the Σ^- and π^+ particles is appreciably less than that of the K⁻ and proton.)

Of the four particles involved in the reaction: K⁻, p, Σ^-, and π^+, only the proton p is stable. Among the other three, the Σ^- is the most unstable, as it lives the shortest time after it is created. After about 10^{-10} seconds a Σ^- particle decays into a π^- meson and a neutral particle (which leaves no track in the bubble chamber). We symbolically write the *sigma decay reaction* as

$$\Sigma^- \rightarrow \pi^- + X^0,$$

where X^0 designates the unseen neutral particle. We shall refer to a particular example of the two reactions, sigma production followed by sigma decay, as an *event*. In Figure 5.14a, we have one such event which has been circled. For clarity, the circled event is shown schematically in Figure 5.14b with the identity and direction of each track indicated. Note how the Σ^- and π^+ tracks appear to be one continuous track, implying that the sigma production reaction took place after the K⁻ came to rest (at point 1). Notice also the kink in the Σ^- track less than 1.0 cm from the point of its creation. It is at the kink (point 2) that the Σ^- decayed into a π^- meson and the unseen neutral particle X^0.

The details of an event, such as the length, direction, and curvature of each track, generally vary from event to event. However, an event can usually be identified by its general configuration, or *topology*. Identifications based on

Figure 5.14   An example of an event

(a)

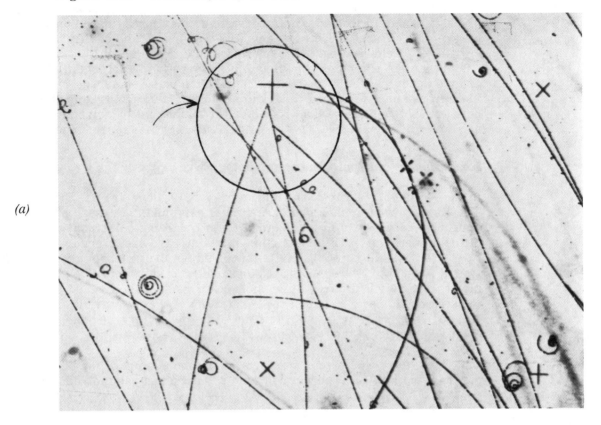

(b)

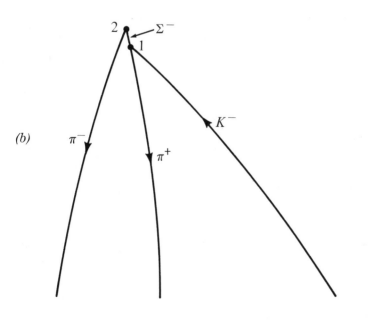

event topology can be confirmed if we make measurements of the length, direction, and curvature of each track, and then analyze these data by computer. Through such a procedure we can determine the identity of the unseen neutral particle X^0. We use the measurements to determine the energy and momentum of both of the charged particles present in the sigma decay (Σ^- and π^-) and then use the laws of energy and momentum conservation to determine the energy and momentum of the X^0 particle. Finally, the mass of X^0, and therefore its identity, can be determined if we know both its energy and momentum. In the next section, we discuss the procedure for making measurements of each track and describe how the X^0 mass is calculated from these measurements.

Measurement of Bubble Chamber Photographs

In each of the photographs in Figures 5.15-5.19 there are one or more events. The circled event in each photograph is of particular interest, because all the tracks lie very nearly in the plane of the photograph, which considerably simplifies the analysis. This is, of course, a very rare occurrence, since all directions are possible for the particles involved. In the general case, we must analyze at least two* stereoscopic photographs of each event to be able to completely describe it in three dimensions. For each of the circled events, we need to determine three quantities: the momentum of the π^- particle (p_π), the momentum of the Σ^- particle (p_Σ) at the point of its decay, and the angle θ between the π^- and Σ^- tracks at the point of decay.

Determination of p_π. We can determine the momentum of the π^- particle if we measure the radius of curvature of the π^- track, and use the relation

$$p = qRB. \tag{5.20}$$

In our system of units, this can be written

$$p = 6.86\ R \tag{5.21}$$

with R in units of cm. (The units for p will be discussed in the next section.) We can determine the radius of the π^- track R, using the *sagitta* method (see Figure 5.20). The Pythagorean theorem gives the following relation between the radius R, chord length ℓ, and sagitta s:

$$R^2 = (R - s)^2 + (\tfrac{1}{2}\ell)^2$$

which can be solved for R to obtain

$$R = \frac{\ell^2}{8s} + \tfrac{1}{2}\,s. \tag{5.22}$$

Thus, we can find the radius R by drawing the longest** possible chord on the track, measuring the chord length ℓ and the sagitta s, and then applying equation 5.22.

---

* In practice three photographs are often used in order to overdetermine the results.

** We want to draw the longest possible chord in order to obtain the greatest accuracy of measurement for the chord length ℓ and sagitta s.

Figure 5.15

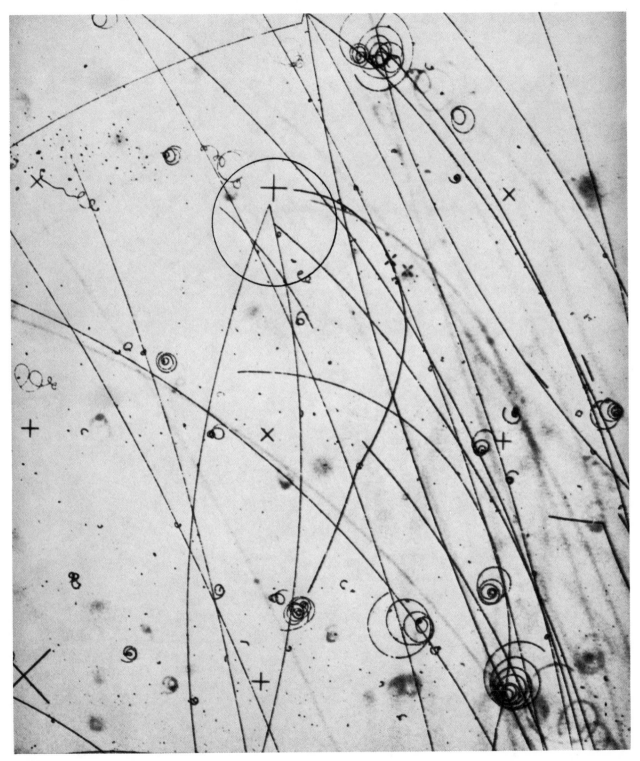

Figure 5.16

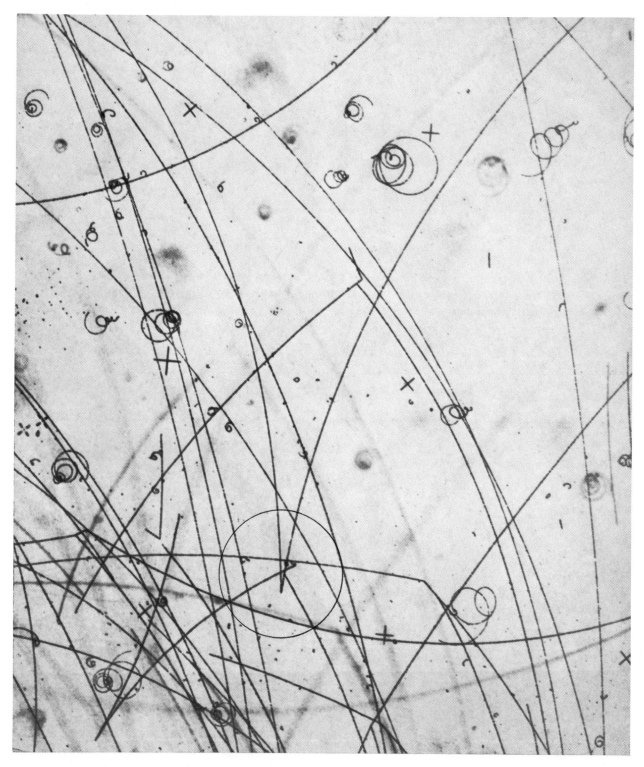

Figure 5.17

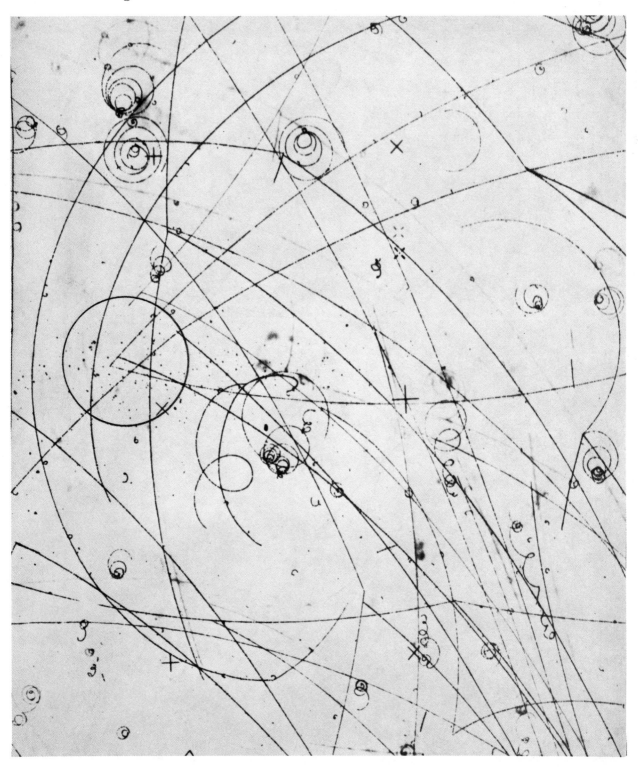

Figure 5.18

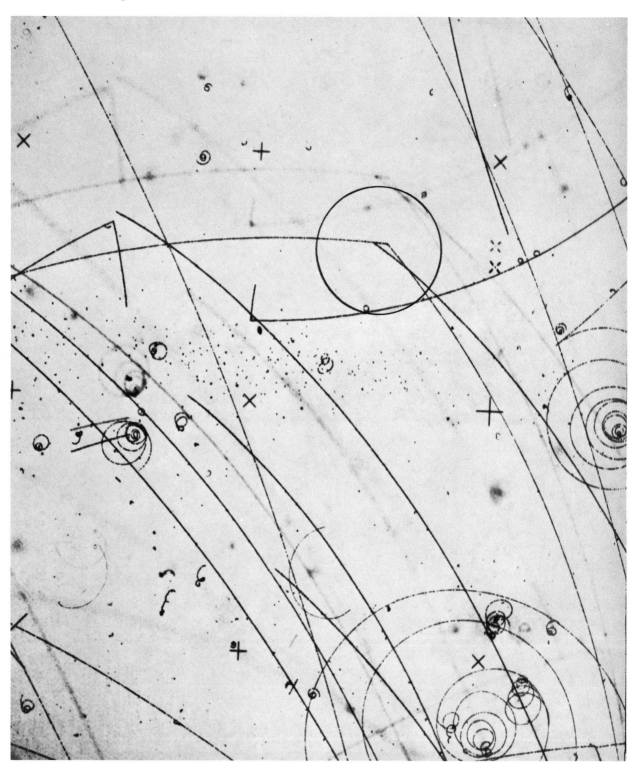

Figure 5.19

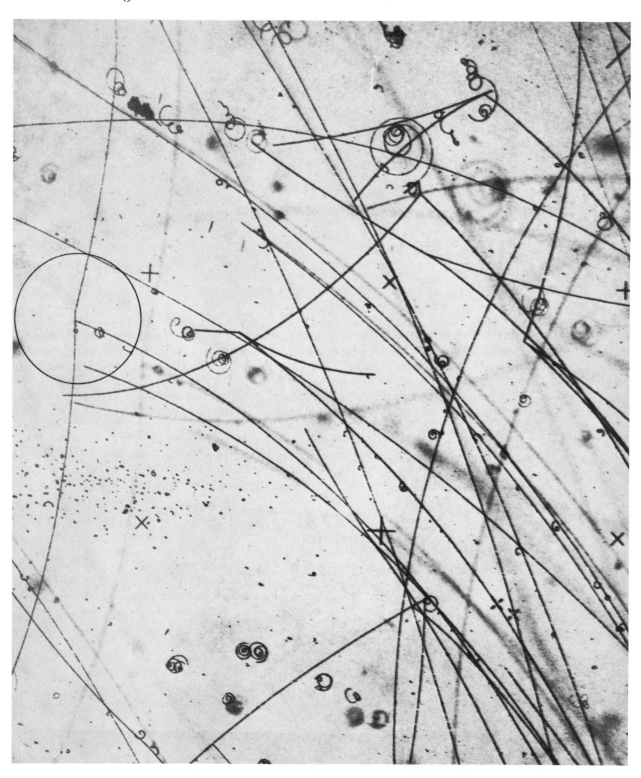

Figure 5.20   Sagitta method for determining radius of curvature

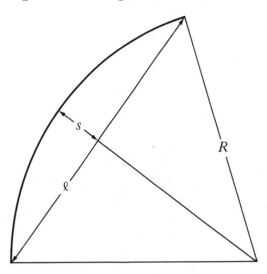

Determination of p_Σ. We cannot determine the momentum of the Σ^- track from its radius of curvature, since the track is much too short. Instead, we make use of the known way that a particle loses momentum as a function of the distance it travels. In each event in which the K^- comes to rest before interacting, momentum and energy conservation applied to the Σ^- decay process requires the Σ^- particle to have a specific momentum of 174 MeV/c. The Σ^- particle, being relatively massive, loses energy rapidly, so that its momentum at the point of its decay is appreciably less than 174 MeV/c, even though it travels only a short distance.

It is known that the distance traveled by a charged particle before it comes to rest (its *range* d) is approximately proportional to the fourth power of its initial momentum, i.e., $d \propto p^4$. For a Σ^- particle traveling in liquid hydrogen, the constant of proportionality is such that a particle of momentum 174 MeV/c has a range of 0.597 cm[*], i.e.,

$$d = 0.597 \left(\frac{p_\Sigma}{174}\right)^4 .$$

In most cases, the Σ^- particle decays *before* it comes to rest, so that the distance it travels, the Σ^- track length ℓ_Σ, is less than the range $d_0 = 0.597$ cm. Using the *residual range* $d_0 - \ell_\Sigma$, we can find p_Σ, the momentum of the Σ^- particle at the point of decay from

$$d_0 - \ell_\Sigma = 0.597 \left(\frac{p_\Sigma}{174}\right)^4 ,$$

which can be solved for p_Σ to yield

$$p_\Sigma = 174 \left(1 - \frac{\ell_\Sigma}{0.597}\right)^{1/4} . \tag{5.23}$$

Note that for $\ell_\Sigma = 0.597$, we have $p_\Sigma = 0$, as expected.

---

[*] This is the range of a Σ^- particle in the photographs in Figures 5.15-5.19 whic are less than life size.

Determination of θ. The angle θ between the π^- and Σ^- momentum vectors can be directly determined if we draw tangents to the π^- and Σ^- tracks at the point of the Σ^- decay. We can then measure the angle between the tangents using a protractor. In Figure 5.21 an alternative method which does not require a protractor is shown. Let AC and BC be the tangents to the π^- and Σ^- tracks respectively. Drop a perpendicular (AB) and measure the distances AB and BC. The ratio AB/BC gives the tangent of the angle $180° - \theta$.

Figure 5.21   Determination of the angle θ

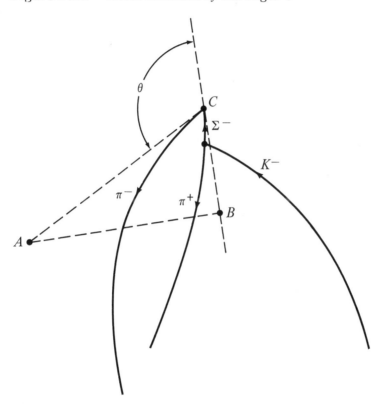

Calculation of the Mass of the X⁰ Particle

The mass of the X^0 particle can be calculated from the four measured quantities: chord length of the π^- track (l_π), sagitta of the π^- track (s), chord length of the Σ^- track* (l_Σ), and the angle between the π^- and Σ^- tracks (θ). The momentum of the π^- track can be calculated from s and l_π using equations 5.22 and 5.21, and the momentum of the Σ^- track can be found from l_Σ using equation 5.23. In our system of units we have energy in MeV, momentum in MeV/c, and mass in MeV/c², where c is the speed of light. Since the particle speeds are all a significant fraction of the speed of light, we must use relativistic formulas. In the above system of units, the relativistic relation between mass and energy becomes**

---

* The Σ^- track is so short that we make a negligible error by using its measured chord length in place of its track length in equation 5.23.

** No factors of c appear in the relativistic formulas as a result of our choice of units.

$$E = m,$$

and the relation between energy and momentum becomes

$$E = \sqrt{p^2 + m_0^2},\tag{5.24}$$

where m_0 is the rest mass. We can find the energy of the π^- and Σ^- particles (E_π and E_Σ) with the aid of equation 5.24, using the experimentally determined momenta and the known rest masses ($m_\Sigma = 1197$ MeV/c^2 and $m_\pi = 140$ MeV/c^2).

We can find the momentum of the X^0 particle by applying the Law of Conservation of Momentum to the Σ^- decay process:

$$\mathbf{p}_\Sigma = \mathbf{p}_\pi + \mathbf{p}_X$$

which can be solved for p_X to obtain

$$\mathbf{p}_X = \mathbf{p}_\Sigma - \mathbf{p}_\pi.\tag{5.25}$$

Similarly, we can find the energy of the X^0 particle by applying the Law of Conservation of Mass-Energy:

$$E_X = E_\Sigma - E_\pi\tag{5.26}$$

If we solve equation 5.24 for m_0, we find

$$m_0 = \sqrt{E^2 - p^2},\tag{5.27}$$

so that we can determine the rest mass of the X^0 particle using

$$m_X = \sqrt{(E_\Sigma - E_\pi)^2 - (\mathbf{p}_\Sigma - \mathbf{p}_\pi)^2},\tag{5.28}$$

which can finally be written as

$$m_X = \sqrt{(E_\Sigma - E_\pi)^2 - p_\Sigma^2 - p_\pi^2 + 2p_\Sigma p_\pi \cos\theta}.\tag{5.29}$$

Using equation 5.29, we have a way of calculating the mass of the unseen neutral particle X^0, in terms of quantities that can be determined from measurements made on the bubble chamber photograph.

Computer Program to Calculate the X^0 Mass

The program given in flow diagram form in Figure 5.22, and listed on pages 256-257, first reads values for the quantities N, LSIG, TH, S, and LPI where

N	=	an arbitrary identifying number for the event
LSIG	=	ℓ_Σ (chord length of the Σ^- track, in cm)
TH	=	θ (angle between π^- and Σ^- tracks, in degrees)
S	=	s (sagitta of π^- track, in cm)
LPI	=	ℓ_π (chord length of π^- track, in cm).

The program then calculates the mass of the X^0 particle.

Figure 5.22   *Flow diagram for calculation of X^0 Mass*

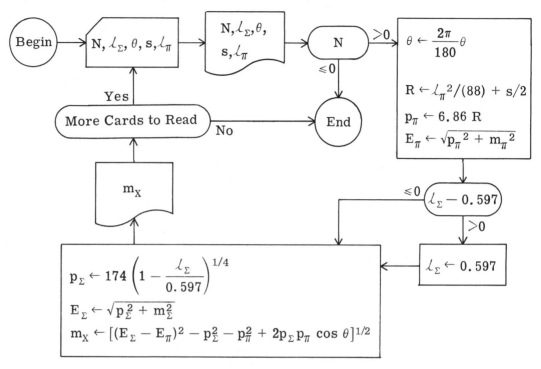

Figure 5.23

```
EVENT NUMBER        1.00

SIGMA TRACK LENGTH=     .50

THETA=   142.0

SAGITTA OF PI=    .970

CHORD LENGTH OF PI=  14.800


MASS OF X PARTICLE=  915.0
```

For the circled event in Figure 5.15, you should obtain values for the parameters close to the following:

N	LSIG	TH	S	LPI
1.0	0.5	142.0	0.97	14.8

A data card containing these values produced the output shown in Figure 5.23.

Problems 6-8 at the end of this chapter pertain to the analysis of measurements from bubble chamber photographs.

```
C                 ANALYSIS OF MEASUREMENTS FROM BUBBLE CHAMBER PHOTOGRAPHS
C
C■-------------------------------------------------------------------------■
C
C
C           THIS PROGRAM ANALYZES MEASUREMENTS MADE ON A SERIES OF BUBBLE
C           CHAMBER PHOTOGRAPHS.  USING MEASUREMENTS MADE OF THE CHARGED
C           TRACKS, THE PROGRAM COMPUTES THE MASS OF AN UNSEEN NEUTRAL
C           PARTICLE  X, WHICH IS PRESENT IN SIGMA DECAY.
C
C
      REAL N,LSIG,LPI
C
C           READ A DATA CARD CONTAINING N, LSIG, TH, S, LPI.
C           N IS THE IDENTIFYING NUMBER FOR AN EVENT.
C           LSIG IS THE LENGTH OF THE SIGMA TRACK.
C           TH IS THE ANGLE BETWEEN THE SIGMA AND PI MINUS TRACKS, IN DEGREES.
C           S IS THE SAGITTA OF THE PI MINUS TRACK, IN CENTIMETERS.
C           LPI IS THE CHORD LENGTH OF THE PI MINUS TRACK.
C
   10 READ(2,1000)N,LSIG,TH,S,LPI
      WRITE(3,1001)N,LSIG,TH,S,LPI
C
C           IF THE DATA CARD READ IN IS BLANK, THEN N IS ZERO AND WE QUIT.
C
      IF(N)30,30,20
   20 CONTINUE
C
C           CONVERT THETA FROM DEGREES TO RADIANS.
C
      THETA=TH*3.14159265/180.
C
C           R IS THE RADIUS OF CURVATURE OF THE PI MINUS TRACK.
C
      R=LPI**2/(8.0*S)+S/2.0
C
C           PPI IS THE MOMENTUM IN MEV/C OF THE PI MINUS TRACK.
C
      PPI=6.86*R
C
C           EPI IS THE TOTAL RELATIVISTIC ENERGY OF THE PI MINUS.
C
      EPI=SQRT(PPI**2+140.0**2)
C
C           THE LONGEST POSSIBLE SIGMA TRACK LENGTH IS 0.597 CM.
C
      IF(LSIG-0.597)24,24,22
   22 LSIG=0.597
   24 CONTINUE
C
C           COMPUTE SIGMA MOMENTUM FROM TRACK LENGTH.
```

```
C
      PSIG=174.0*(1.0-LSIG/0.597)**0.25
C
C           ESIG IS THE TOTAL RELATIVISTIC ENERGY OF THE SIGMA MINUS.
C
      ESIG=SQRT(PSIG**2+1197.0**2)
C
C           COMPUTE THE MASS OF THE X PARTICLE.
C
      X=SQRT((ESIG-EPI)**2-PSIG**2-PPI**2+2.0*PSIG*PPI*COS(THETA))
      WRITE(3,1002)X
C
C           READ IN DATA FOR THE NEXT EVENT.
C
      GO TO 10
 1000 FORMAT(5F10.5)
 1001 FORMAT(20X,13H EVENT NUMBER,F10.2//20X,20H SIGMA TRACK LENGTH=,
     1F7.2//20X,7H THETA=,F7.1//20X,15H SAGITTA OF PI=,F7.3//20X,
     2 20H CHORD LENGTH OF PI=,F8.3///)
 1002 FORMAT(20X,20H MASS OF X PARTICLE=,F7.1//////)
   30 CONTINUE
      CALL EXIT
      END
```

Problems for Chapter 5

Least Squares Fitting of Experimental Data

1. Least squares straight line fit.   Run the program using data obtained from some experiment. (A suggested experiment is mentioned below.) The measured quantities from the experiment should satisfy a linear relation, or they should be so related that they can be transformed into a pair of linearly related variables, as in the example on page 230. Discuss the output you obtain using the least squares straight line fitting program, and comment on the goodness of the fit and the evidence for or against the presence of systematic errors in the data.

One simple experiment involves rolling a die 150 times. For each of the numbers $1, 2, \ldots, 6$, record how many times that number comes up: $n_1, n_2, \ldots, n_6$. Let the numbers $1, 2, \ldots, 6$ be the x-values, and the numbers $n_1, n_2, \ldots, n_6$ be the y-values. If the die is fair, we expect the data to be consistent with a straight line $y = \alpha_1 + \alpha_2 x$, with y-intercept, $\alpha_1 = 25$ and slope $\alpha_2 = 0$. According to the laws of statistics, the measurement errors in this case are given by $\sqrt{n_1}, \sqrt{n_2}, \sqrt{n_3}, \ldots, \sqrt{n_6}$.

2. One-parameter least squares straight line fit.   Sometimes we want a straight line fit in which only one parameter is adjustable and the other is fixed. For example, if the y-intercept is fixed at zero, then the least squares fit straight line has a slope α_2, given by

$$\alpha_2 = \frac{S_{xy}}{S_{xx}} = \left(\sum_j \frac{x_i y_i}{\Delta y_j{}^2} \right) \bigg/ \left(\sum_j \frac{x_j{}^2}{\Delta y_j{}^2} \right)$$

and an uncertainty given by

$$\Delta \alpha_2 = \frac{1}{\sqrt{S_{XX}}} = \left(\sum_j \frac{x_j^2}{\Delta y_j^2} \right)^{-1/2} .$$

This type of one-parameter fit is applicable to the data from the air track experiment, since it is known a priori that the y-intercept should be zero in that case.

Another example of a one-parameter straight line fit is a straight line of zero slope and adjustable y-intercept. In this case (page 227), the least squares fit straight line has a y-intercept given by

$$\alpha_1 = \left(\sum y_j/\Delta y_j^2 \right) \Big/ \left(\sum \frac{1}{\Delta y_j^2} \right)$$

and an uncertainty given by

$$\Delta \alpha_1 = \frac{1}{\left(\sum_j \frac{1}{\Delta y_j^2} \right)^{1/2}} .$$

This is applicable to the data from the die throwing experiment mentioned in problem 1.

Modify the least squares straight line fitting program so that it computes and plots the least squares fit for the two types of one-parameter straight line fits just discussed. Run the modified version of the program and discuss the results.

3. *Relation between χ^2 and probability.*   The probability that the value of χ^2 computed for a frequency distribution having n degrees of freedom exceeds a value χ_0^2 is given by

$$P(\chi_0^2, n) = \int_0^{\chi_0^2} t^{(n/2)-1} e^{-t/2} \, dt \Big/ \int_0^\infty t^{(n/2)-1} e^{-t/2} \, dt \qquad (5.30)$$

This result assumes that the theoretical frequency distribution used to compute χ^2 is the correct one, and that the measured quantity is subject to Gaussian distributed errors of a known standard deviation.

Write a program to evaluate the probability for various values of χ_0^2 and n using equation 5.30 by evaluating the integrals using some numerical method. Check your results against the values in Table 5.1.

4. *Finding the minimum χ^2 by a search procedure.*   Write a program to explicitly calculate and print out χ^2 for a series of values of α_1 and α_2 for each of the one-parameter fits discussed in problem 2. In this way you can empirically find those values of the parameters which give the smallest value of χ^2, and thereby verify that the equations given in problem 2 for the least squares fit are correct. You may wish to extend this so that the program does an automatic search, using the decision tree algorithm discussed in Chapter 1, which closes in on the minimum value of χ^2.

5. *Nonlinear least squares two-parameter fit.* A two-parameter least squares fit to an arbitrary function involving two adjustable parameters can be made by finding the value of χ^2 at a two-dimensional grid of values for α_1 and α_2. If we display the χ^2 values at all points on the grid using a mapping technique, such as the one used in the equipotential plotting problem in Chapter 2, we can obtain a set of contours of constant χ^2 in the $\alpha_1\alpha_2$-plane. From such contours, we can determine the values of α_1 and α_2 for which χ^2 is a minimum.

Write a program to carry out such a procedure and run the program using a set of data. If you use the data listed in Figure 5.10, you will be able to compare the exact results from the straight line least squares fit with the approximate values you obtain using the graphical method.

An alternative procedure to find the values of α_1 and α_2 for which χ^2 is a minimum is to use the hill-climbing technique discussed in Chapter 1.

Analysis of Measurements from Bubble Chamber Photographs

6. *Calculation of the X^0 particle's mass.* Make measurements on each of photographs in Figures 5.15-5.19 and run the program to calculate the X^0 mass with the data from your measurements. The values on your first data card for the event in Figure 5.15 should be close to those given on page 256. Try to identify the X^0 particle based on the computed masses for each event. The masses of some of the known neutral particles are given below.

particle	mass (in MeV/c^2)
π^0	135
K^0	498
n	940
Λ^0	1116
Σ^0	1192
Ξ^0	1315

7. *Estimate of the error in the X^0 mass.* To estimate the error in the X^0 mass, one simple technique is to vary each of the quantities on the data card by what you judge to be its experimental error, and see what the effect is on the computed X^0 mass. From the most critical quantity (the one that causes the largest shift in the X^0 mass), you can make an estimate of the error in the X^0 mass, and compare this with the spread in values from event to event.

8. *Calculation of the Σ^- lifetime.* The Σ^- lifetime can be approximately determined using the measured values of the Σ^- track lengths. The average momentum of the Σ^- particle can be found from its initial and final values:

$$\bar{p}_\Sigma = \tfrac{1}{2}(173 + p_\Sigma),$$

where p_Σ is found from equation 5.23, using the measured track length ℓ_Σ. The length of time that the Σ^- lives (the time between its creation and decay) is

$$t = \frac{\ell_\Sigma}{v},$$

where ℓ_Σ is the length of the Σ^- track and v is the average velocity of the Σ^- particle. This time interval is measured in a coordinate system at rest in the laboratory. In a coordinate system moving with the particle, the "rest system," the time is given by

$$t = \frac{l_\Sigma}{v} \sqrt{1 - \frac{v^2}{c^2}} \, ,$$

which can be put in the form

$$t = \frac{m \, l_\Sigma}{p_\Sigma} \, ,$$

where m = 1193 is the rest mass of the Σ^- particle.

Write a program that calculates the amount of time that each Σ^- lives, and determine an average lifetime. The accepted value is 1.49×10^{-10} seconds. Since all photographs are less than life-size, the computed times must be multiplied by the scale factor 1.71.

Appendix I

References for Topics in Table 1.1

The topics listed in Table 1.1 are covered in many of the commonly used introductory physics texts. In Table A.1, we find the chapter(s) in each of 24 physics texts which deal with each of the topics. The full title and publisher of each text listed in Table A.1 by author, are given on page 263. As an example of the use of Table A.1, we notice that *Physics* by Alonso and Finn treats the material in topic 3.2 in chapters 23 and 28. As can be seen from Table 1.1, topic 3.2 deals with waves in two dimensions, in particular, the Doppler Effect, shock waves, and interference.

Table 1.1

Physics Topics	Associated Mathematical and/or Computer Methods
2.1 Equipotential Surfaces for Two Point Charges	Computer Graphics and Intensity Scaling
2.2 Potential for a Charged Thin Wire	Numerical Integration by Trapezoidal and Simpson's Rules and Interpolating Polynomials
2.3 Discharge of a Capacitor in an RC Circuit	Numerical Solution of Differential Equations Using Euler's Method
2.4 The RLC Series Circuit	Numerical Solution of Differential Equations Using *Improved* Euler's Method
3.1 Superposition of Waves	Fourier Analysis and Synthesis of Wave Forms
3.2 Waves in Two Dimensions: Doppler Effect, Shock Waves, and Interference	Computer Graphics and Superposition of Waves
3.3 Solution of the Schrödinger Equation	Use of Boundary Conditions to Select Proper Solutions: Eigenfunctions and Eigenvalues
4.1 Random Processes	Generation of Random Numbers by Computer: the Power Residue Method
4.2 Center of Mass Coordinates	Numerical Integration Using the Monte Carlo Method
4.3 Fission Chain Reaction: the Critical Mass	Simulation of a Random Process Using the Monte Carlo Method
4.4 The Approach to Equilibrium: the Second Law of Thermodynamics	Simulation of a Random Process Using the Monte Carlo Method
5.1 Fitting of Experimental Data	Least Squares Fitting
5.2 Measurement of Bubble Chamber Photographs	"Missing Mass" Calculation

Table A.1   Chapters in physics texts relating to topics in Table 1.1

Book	Topic number												
	2.1	2.2	2.3	2.4	3.1	3.2	3.3	4.1	4.2	4.3	4.4	5.1	5.2
Alonso and Finn	16	16	21	9, 21	9	23, 28	30		4	22	14		22
Armstrong and King				11		15, 16					19		
Borowitz and Beiser	17	17		26	25	25, 29			8	37			38
Bueche	20	20	21	28		30, 32	37		9	39			40
Carr and Weidner	14	14				18			6	27			28
Feynman, Leighton & Sands I				23	50	29, 51				19		6	
Feynman, Leighton & Sands II	4	4											
French				3, 4	6	2, 8							
Halliday and Resnick	25	25	28		16	16, 17				12			
Kip	3	3			11								
McCormick	33	33		47	30	30, 52				17	59		58
Morgan I						24			7				
Morgan II	25	25	31	30		34							
Purcell	2	2	4	8									
Reimann I					11	3			7		18		
Reimann II	20	20	25	33		38							
Resnick & Halliday I					19	19, 20			14		25		
Resnick & Halliday II		29	32	38		40, 43							
Sears and Zemansky	26	26	29	33, 35		41			3	44			
Tipler							6			10	2		10
Weidner and Sells I									8		22		
Weidner and Sells II	28	28	30	37	39	45							
Weidner and Sells III							5			10			8, 11
Young					3, 6	2, 3	6, 7			12			13

Titles and Publishers of the Textbooks Listed by Author in Table A.1

Alonso, M., and Finn, E. 1970. Physics. Reading, Mass.: Addison-Wesley Publishing Co.

Armstrong, R., and King, J. 1970. Mechanics, Waves and Thermal Physics. Englewood Cliffs, N.J.: Prentice-Hall.

Borowitz, S., and Beiser, A. 1971. Essentials of Physics. Reading, Mass.: Addison-Wesley Publishing Co.

Bueche, F. 1969. Introduction to Physics for Scientists and Engineers. New York: McGraw-Hill Book Co.

Carr, H., and Weidner, R. 1971. Physics from the Ground Up. New York: McGraw-Hill Book Co.

Feynman, R., Leighton, R., and Sands, M. 1964. The Feynman Lectures on Physics, Vol. I. Reading, Mass.: Addison-Wesley Publishing Co.

Feynman, R., Leighton, R., and Sands, M. 1964. The Feynman Lectures on Physics, Vol. II. Reading, Mass.: Addison-Wesley Publishing Co.

French, A. 1971. Vibrations and Waves. New York: W.W. Norton and Co.

Halliday, D. and Resnick, R. 1970. Fundamentals of Physics. New York: John Wiley and Sons.

Kip, A. 1969. Fundamentals of Electricity and Magnetism. New York: McGraw-Hill Book Co.

McCormick, W. 1969. Fundamentals of University Physics. Toronto: Macmillan Co.

Morgan, J. 1969. Introduction to University Physics, Vol. I. Boston: Allyn and Bacon.

Morgan, J. Introduction to University Physics, Vol. II. Boston: Allyn and Bacon.

Purcell, E. 1965. Electricity and Magnetism, Berkeley Physics Course, Vol. II. New York: McGraw-Hill Book Co.

Reimann, A. 1971. Volume I: Mechanics and Heat. New York: Barnes and Noble.

Reimann, A. 1971. Volume II: Electricity and Magnetism. New York: Barnes and Noble.

Resnick, R. and Halliday, D. 1962. Physics for Students of Science and Engineering, Vol. I. New York: John Wiley & Sons.

Resnick, R. and Halliday, D. 1962. Physics for Students of Science and Engineering, Vol. II. New York: John Wiley & Sons.

Sears, F., and Zemansky, M. 1970. University Physics. Reading, Mass.: Addison Wesley Publishing Co.

Tipler, P. 1969. Foundations of Modern Physics. New York: Worth Publishers.

Weidner, R., and Sells, R. 1965. Elementary Classical Physics, Vol. I. Boston: Allyn and Bacon.

Weidner, R., and Sells, R. 1965. Elementary Classical Physics, Vol. II. Boston: Allyn and Bacon.

Weidner, R., and Sells, R. 1968. Elementary Modern Physics. Boston: Allyn and Bacon.

Young, H. 1968. Fundamentals of Optics and Modern Physics. New York: McGraw-Hill Book Co.

Appendix II

Computer-Dependent Features of Programs

The programs in this book are written in the FORTRAN IV language and have been run on an 8,000 word, 16 bit, IBM 1130 computer. They should run on any computer which uses FORTRAN IV with little modification. Possible modifications include the following.

1. Numbers for input-output devices. Since the numbers assigned to various input-output devices vary with different computers, these may have to be changed in the programs. At present, it is assumed that input to the program is from device number 2 (the card reader on the IBM 1130), and output is to device number 3 (the printer on the IBM 1130).

2. Rescaling of graphs. Many of the programs generate graphs which print up to 128 characters on a line. If the printer is not capable of printing this many characters on a line, the graphs will need to be rescaled.

3. Use of quotes to represent text. The DATA statements appearing in many of the programs use single quotes to designate alphanumeric data (text), instead of Hollerith (H) format (see page 47), which may be required on some computers.

4. Changes in the random number generator. The technique used to generate random numbers (discussed on pages 178-180) should work on most binary computers which can represent integers as large as 32767. The technique would have to be modified for a decimal computer, or for one on which the largest integer that can be represented is less than 32767. A modification would also be required for a computer which does not give the least significant part of the result when the multiplication of two fixed point numbers results in an overflow.

Appendix III

Determination of
Fourier Coefficients a_k and b_k

The Fourier expansion:

$$F(x) = \sum_{k=0}^{\infty} a_k \cos kx + \sum_{k=1}^{\infty} b_k \sin kx \qquad (A.1)$$

can be used to derive the formulas for the coefficients a_k and b_k:

$$a_k = \frac{1}{\pi} \int_{-\pi}^{+\pi} F(x) \cos kx \, dx \quad \text{for } k = 1, 2, 3, \ldots \qquad (A.2)$$

$$b_k = \frac{1}{\pi} \int_{-\pi}^{+\pi} F(x) \sin kx \, dx \quad \text{for } k = 1, 2, 3 \ldots \qquad (A.3)$$

To derive A.2, we multiply both sides of equation A.1 by the function $\cos mx$ and then integrate both sides between $x = -\pi$ and $x = +\pi$; interchanging the order of the integration and summation we obtain

$$\int_{-\pi}^{+\pi} F(x) \cos mx \, dx = \sum_{k=0}^{\infty} a_k \int_{-\pi}^{+\pi} \cos mx \cdot \cos kx \, dx$$

$$+ \sum_{k=1}^{\infty} b_k \int_{-\pi}^{+\pi} \cos mx \cdot \sin kx \, dx \qquad (A.4)$$

Evaluating the integrals appearing inside the sums on the right-hand side of equation A.4, we find

$$\int_{-\pi}^{+\pi} \cos mx \cdot \cos kx \, dx = \begin{cases} \pi & \text{for } m = k \\ 0 & \text{for } m \neq k \end{cases}$$

and

$$\int_{-\pi}^{+\pi} \cos mx \cdot \sin kx \, dx = 0 \qquad \text{for all } m \text{ and } k.$$

Therefore, the only nonzero term in either sum on the right-hand side of equation A.4 is that involving a_m, so that we have

$$\int_{-\pi}^{+\pi} F(x) \cos mx \, dx = \pi a_m.$$

We can then obtain equation A.2 by simply replacing m by k in the above equation.

We can derive equation A.3 in a similar manner, if we multiply both sides of equation A.1 by sin mx and integrate both sides between $x = -\pi$ and $x = +\pi$:

$$\int_{-\pi}^{+\pi} F(x) \sin mx \, dx = \sum_{k=0}^{\infty} a_k \int_{-\pi}^{+\pi} \sin mx \cdot \cos kx \, dx$$

$$+ \sum_{k=1}^{\infty} b_k \int_{-\pi}^{+\pi} \sin mx \cdot \sin kx \, dx \qquad (A.5)$$

Evaluating the integrals appearing inside the sums on the right-hand side of equation A.5, we find

$$\int_{-\pi}^{+\pi} \sin mx \cdot \cos kx \, dx = 0 \qquad \text{for all m and k}$$

and

$$\int_{-\pi}^{+\pi} \sin mx \cdot \sin kx \, dx = \begin{cases} \pi & \text{for } m = k \\ 0 & \text{for } m \ne k. \end{cases}$$

Therefore, the only nonzero term in either sum on the right-hand side of equation A.5 is that involving b_m, so that we have

$$\int_{-\pi}^{+\pi} F(x) \sin mx \, dx = \pi b_m.$$

We can then obtain equation A.3 by replacing m by k in the above equation.

Appendix IV

Fourier Coefficients for Square, Triangular, and Spiked Wave Forms

For the square wave, we have

$$a_k = \frac{1}{\pi} \int_{-\pi}^{+\pi} F(x) \cos kx\, dx = 0 \qquad \text{for } k = 1, 2, 3, \ldots .$$

The integral vanishes because the integrand is an antisymmetric* function, being the product of an antisymmetric function ($F(x)$) and a symmetric function ($\cos kx$). The b_k coefficients for the square wave can be found using

$$b_k = \frac{1}{\pi} \int_{-\pi}^{+\pi} F(x) \sin kx\, dx.$$

With the definition of the square wave: ($F(x) = +1$ for $0 \leq x < \pi$; $F(x) = -1$ for $-\pi \leq x < 0$), this becomes

$$b_k = -\frac{1}{\pi} \int_{-\pi}^{0} \sin kx\, dx + \frac{1}{\pi} \int_{0}^{\pi} \sin kx\, dx.$$

These integrals can be easily evaluated, giving

$$\text{square wave: } b_k = \frac{2}{\pi k} (1 - (-1)^k) \qquad \text{for } k = 1, 2, 3, \ldots .$$

In the case of the triangular wave, we find

$$b_k = \frac{1}{\pi} \int_{-\pi}^{\pi} F(x) \sin kx\, dx = 0 \quad (\text{why?}) .$$

The a_k coefficients for the triangular wave can be found using

$$a_k = \frac{1}{\pi} \int_{-\pi}^{\pi} F(x) \cos kx\, dx.$$

With the definition of the triangular wave: $\left(F(x) = 1 - \frac{2}{\pi} |x| \right)$, this becomes

$$a_k = \frac{1}{\pi} \int_{-\pi}^{0} \left(1 + \frac{2x}{\pi} \right) \cos kx\, dx + \frac{1}{\pi} \int_{0}^{+\pi} \left(1 - \frac{2x}{\pi} \right) \cos kx\, dx.$$

These two integrals can be easily evaluated, giving

$$\text{triangular wave: } a_k = \left(\frac{2}{\pi k} \right)^2 (1 - (-1)^k) \qquad \text{for } k = 1, 2, 3, \ldots .$$

---

* $F(x)$ is symmetric if $F(-x) = F(x)$. It is antisymmetric if $F(-x) = -F(x)$.

Finally, in the case of the spiked wave, we have

$$b_k = \frac{1}{\pi} \int_{-\pi}^{+\pi} F(x) \sin kx \, dx = 0 \quad (why?).$$

The coefficient a_k for the spiked wave can be found using

$$a_k = \frac{1}{\pi} \int_{-\pi}^{\pi} F(x) \cos kx \, dx.$$

With the definition of the spiked wave: $\left(F(x) = \frac{1}{\epsilon} \text{ for } |x| < \frac{\epsilon}{2} ; F(x) = 0 \text{ for } |x| > \frac{\epsilon}{2} \right)$, this can be written

$$a_k = \frac{1}{\pi} \int_{-\epsilon/2}^{+\epsilon/2} \frac{1}{\epsilon} \cos kx \, dx.$$

Since we are assuming ϵ to be infinitesimally small for the spiked wave, we have $\cos kx \approx 1$, for all x in the interval $-\frac{\epsilon}{2} < x < +\frac{\epsilon}{2}$. Thus, we obtain

$$\text{spiked wave: } a_k = \frac{1}{\pi} \quad \text{for } k = 1, 2, 3, \ldots.$$

Note that it is only for the spiked wave that the a_k coefficients are independent of k, i.e., they are constant. The particular value of the constant does not affect the shape of the wave form, but only the scale. We shall therefore choose a different constant than $\frac{1}{\pi}$ to obtain a more convenient scale. We observe that for the spiked wave form, the Fourier synthesized wave form involving N waves, $F_N(x)$, has unit height at $x = 0$, provided $a_k = \frac{1}{N}$, for $k = 1, 2, 3, \ldots, N$. We shall use this constant.

Appendix V

Determination of Position
of a Moving Source

We wish to determine the position $(x_0, 0)$ of a moving source at the time when the wave front reaching an observer at (x, y) and $t = 0$ was being emitted from the source $(t = -T)$. As can be seen in Figure 3.23, the distance traveled by the wave to reach the point (x, y) from the point $(x_0, 0)$ is

$$r = cT = ((x - x_0)^2 + y^2)^{1/2},$$

where c is the *wave* velocity. If the source is located at a known point $(D, 0)$ at $t = 0$, then we may write

$$D - x_0 = vT,$$

where v is the *source* velocity. Eliminating T between the above pair of equations, we obtain

$$\frac{c}{v} (D - x_0) = ((x - x_0)^2 + y^2)^{1/2}.$$

With the aid of a little algebra, the above equation can be written as a quadratic equation in x_0:

$$\underbrace{\left[1 - \left(\frac{c}{v}\right)^2\right]}_{A} x_0^2 + \underbrace{\left[\frac{2Dc^2}{v^2} - 2x\right]}_{B} x_0 + \underbrace{\left[x^2 + y^2 - \left(\frac{Dc}{v}\right)^2\right]}_{C} = 0$$

Note that no solution of this equation exists if $B^2 - 4AC < 0$; this occurs when $\frac{v}{c} > 1$ and the point (x, y) lies outside the Mach cone. One special case occurs when $\frac{v}{c} = 1$, giving $A = 0$, and the solution is $x_0 = -C/B$. Otherwise, we can solve for x_0 using the quadratic formula:

$$x_0 = \frac{-B + (B^2 - 4AC)^{1/2}}{2A}$$

We reject the other root, since it corresponds to a point which lies to the right of $(D, 0)$ — see Figure 3.23.

Bibliography

The books listed are part of the large body of literature on numerical methods. Most of the books discuss methods that are compatible with computers; some make use of the FORTRAN language.

Acton, Forman S. 1970. Numerical Methods That Work. New York: Harper and Row Publishers.

Arden, Bruce, and Astill, Kenneth. 1970. Numerical Algorithms: Origins and Applications. Reading, Mass.: Addison-Wesley Publishing Co.

Beckett, R. and Hurt, J. 1967. Numerical Calculations and Algorithms. New York: McGraw-Hill Book Co.

Carnahan, Brice; Luther, H. A. and Wilkes, James O. 1969. Applied Numerical Methods. New York: John Wiley and Sons.

Conte, Samuel. 1965. Elementary Numerical Analysis: An Algorithmic Approach. New York: McGraw-Hill Book Co.

Fox, Leslie and Mayers, D. F. 1968. Computing Methods for Scientists and Engineers. New York: Oxford University Press.

Froberg, Carl E. 1969. Introduction to Numerical Analysis. Reading, Mass.: Addison-Wesley Publishing Co.

Gastinel, N. 1969. Linear Numerical Analysis. New York: Academic Press.

Gerald, Curtis F. 1970. Applied Numerical Analysis. Reading, Mass.: Addison-Wesley Publishing Co.

Greenspan, Donald. 1970. Introduction to Numerical Analysis and Applications. Chicago: Markham Publishing Co.

Grove, Wendell E. 1966. Brief Numerical Methods. Englewood Cliffs, N.J.: Prentice-Hall.

Haggerty, Gerald B. 1971. Elementary Numerical Analysis with Programming. Boston: Allyn and Bacon.

Hamming, R. W. 1971. Introduction to Applied Numerical Analysis. New York: McGraw-Hill Book Co.

Hamming, Richard. 1968. Calculus and the Computer Revolution. Boston: Houghton Mifflin Co.

Hamming, Richard W. 1962. Numerical Methods for Scientists and Engineers. New York: McGraw-Hill Book Co.

Handscomb, David C. 1966. Methods of Numerical Approximation. Elmsford, N.Y.: Pergamon Press.

Henrici, Peter K. 1964. Elements of Numerical Analysis. New York: John Wiley and Sons.

Isaacson, Eugene and Keller, H. B. 1966. Analysis of Numerical Methods. New York: John Wiley and Sons.

Jacquez, John A. 1970. A First Course in Computing and Numerical Methods. Reading, Mass.: Addison-Wesley Publishing Co.

Jennings, Walter. 1964. First Course in Numerical Methods. Toronto: Macmillan Co.

Kelly, L. G. 1967. Handbook of Numerical Methods and Applications. Reading, Mass.: Addison-Wesley Publishing Co.

Khabaza, I. M. 1965. Numerical Analysis. Elmsford, N.Y.: Pergamon Press.

Kunz, Kaiser L. 1957. Numerical Analysis. New York: McGraw-Hill Book Co.

Kuo, Shan S. 1965. Numerical Methods and Computers. Reading, Mass.: Addison-Wesley Publishing Co.

Lee, John A. 1966. Numerical Analysis for Computers. New York: Van Nostrand Reinhold Co.

Lieberstein, H. 1968. Course in Numerical Analysis. New York: Harper and Row Publishers.

McCalla, Thomas R. 1967. Introduction to Numerical Analysis and FORTRAN Programming. New York: John Wiley and Sons.

McCormick, John M., and Salvadori, M. G. 1964. Numerical Methods in Fortran. Englewood Cliffs, N.J.: Prentice-Hall.

McCracken, Daniel D. and Dorn, William S. 1964. Numerical Methods and Fortran Programming. New York: John Wiley and Sons.

Merritt, Frederick S. 1970. Modern Mathematical Methods in Engineering. New York: McGraw-Hill Book Co.

Moursand, David G. and Duris, C. S. 1967. Elementary Theory and Application of Numerical Analysis. New York: McGraw-Hill Book Co.

Nielsen, Kaj L. 1964. Methods in Numerical Analysis. Toronto: Macmillan Co.

Norkin, Sim B. 1965. Elements of Computational Mathematics. Elmsford, N.Y.: Pergamon Press.

Pettofrezzo, Anthony J. 1967. Introductory Numerical Analysis. Lexington, Mass.: D. C. Heath and Co.

Price, W. T. and Miller, M. 1967. Elements of Data Processing Mathematics. New York: Holt, Rinehart and Winston.

Ralston, Anthony. 1965. First Course in Numerical Analysis. New York: McGraw-Hill Book Co.

Redish, K. A. 1962. Introduction to Computational Mathematics. New York: John Wiley and Sons.

Salvadori, Mario G. and Baron, M. L. 1961. Numerical Methods in Engineering. Englewood Cliffs, N.J.: Prentice-Hall.

Scarborough, James B. 1966. Numerical Mathematical Analysis. Baltimore, Md.: Johns Hopkins Press.

Scheid, Francis. 1968. Numerical Analysis. (Schaum's Outline Series). New York: McGraw-Hill Book Co.

Singer, James. 1967. Elements of Numerical Analysis. New York: Academic Press.

Southworth, R. and De Leeuw, S. 1965. Digital Computation and Numerical Methods. New York: McGraw-Hill Book Co.

Stanton, Ralph G. 1961. Numerical Methods for Science and Engineering. Englewood Cliffs, N.J.: Prentice-Hall.

Stark, Peter A. 1970. Introduction to Numerical Methods. Toronto: Macmillan Co.

Stiefel, E. L. 1963. Introduction to Numerical Mathematics. Translated by W. C. Rheinboldt. New York: Academic Press.

Tompkins, Charles B. and Wilson, Walter L. Jr. 1969. Elementary Numerical Analysis. Englewood Cliffs, N.J.: Prentice-Hall

Watson, W. A., et al. 1969. Numerical Analysis: The Mathematics of Computing, Vol. 1. New York: American Elsevier Publishing Co.

Wendroff, Burton. 1969. First Principles of Numerical Analysis. Reading, Mass.: Addison-Wesley Publishing Co.

Whittaker, Edmund and Robinson, G. 1967. Calculus of Observations: An Introduction to Numerical Analysis. New York: Dover Publications.

Wilkes, Maurice V. Short Introduction to Numerical Analysis. New York: Cambridge University Press.

INDEX